ANIMAL COMMUNITIES

INTERNATIONAL LIBRARY

MICHAEL CHINERY

ANIMAL COMMUNITIES

COLLINS · PUBLISHERS

London · Glasgow

FRANKLIN WATTS, INC.

New York

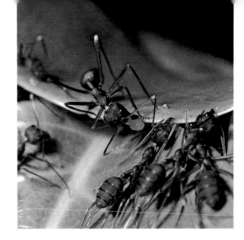

First Edition 1972
Second impression 1975
Third impression 1978

ISBN   0 00   100167   1   *(Collins cased edition)*
ISBN   0 00   103362   X   *(Collins paperback edition)*
SBN   531   02107   6   *(Watts cased edition)*

CONTENTS

INTRODUCTION

A leopard stalks silently towards an antelope and then, with one tremendous leap, brings it crashing to the ground, dead with a broken neck. Quite often the antelope is too large for one meal so the leopard will carry it high up into a tree for safe keeping. The leopard works entirely alone, for it is one of the truly solitary animals. As soon as the leopard cubs are old enough to look after themselves they leave their mothers for good and, except for a very brief period during the mating season, they have nothing to do with other adult leopards again.

In fact, most animals lead solitary lives and have very little contact with other members of the same species, but they are not necessarily quite so alone as the leopard. For example, dozens of small tortoiseshell butterflies will gather on the flower heads of the orpine plant in late summer. They gather there because they are attracted by the rich and abundant nectar in the flowers. They are not associating with each other, and they are probably almost unaware of each other. True, if one butterfly suddenly flies off some of the others may follow, but this is not always so and there is certainly no co-operation between them. The butterflies are solitary insects, each one seeking to satisfy its own needs and no more.

But there are many creatures which do form regular and permanent associations with other members of their own species. These are the *social* creatures, and their associations or communities always have some kind of structure, with each member having a role to play. The animals co-operate for the good of the community as a whole and they are not concerned merely with satisfying their individual needs. Such associations may be limited to the members of a single family, as that of the beavers, or they may involve much larger communities. Some of these communities are rather loosely knit together, good examples being the breeding colonies of various seabirds. On the other hand, we find the very tightly knit communities of the bees and other social insects.

The study of social behaviour in animals has made great advances in recent years and many species previously thought to be more or less solitary are now known to form structured communities of one kind or another. This book describes the various levels of social life found in the animal kingdom and shows how such social behaviour may have evolved from solitary habits.

Zebras live in small groups, consisting of an adult male and up to six mares and their young. One animal often seems to remain on guard while the others drink.

CHAPTER 1

ANIMAL LANGUAGE

Each year, during October, the natives of Samoa and other islands in the South Pacific prepare their boats and nets for a very important event: the annual palolo harvest is drawing near. The palolo is a worm which lives among the offshore corals and, while the Samoans are preparing their fishing gear, the hind end of the worm is swelling with eggs or sperm. Then, at low tide, about a week after the full moon, the hind end breaks free from the rest of the worm and floats to the surface with its cargo of reproductive cells. There are countless millions of palolos on the reefs and they all release their hind portions at the same time. The water becomes like soup, and this is the signal for the Samoans to go out and reap a nutritious harvest. Those palolos that escape the nets release their eggs and sperm into the water. The eggs are fertilized in the water and eventually grow into new worms.

No communication between the worms is necessary to bring about this remarkable swarming behaviour, although there is obviously some chemical attraction between the eggs and sperms after they have been released. The swarming is initiated by the state of the tide and is controlled ultimately by the moon. That is why the islanders can predict exactly when the palolos will appear.

Reproduction in oysters, sea urchins and various other marine creatures is also controlled by the moon and the tides. Although they do not necessarily swarm in the water, the animals release their eggs and sperm as a result of a common reaction to the state of the tide. Such methods are, of course, rather wasteful because only a small proportion of the eggs are ever fertilized.

The need for language

Most animal species have evolved more elaborate reproductive methods involving the pairing of two individuals. This has also necessitated the development of some method of communication between the members of a species, for if two animals are to meet and mate they must be able to find each other and convey their intentions. Even the most solitary of animals must be able to communicate in this way. Mating calls may be the only signals given by some animals, but most species have a much larger repertoire. Those species which look after their young must be able to warn them of danger and to tell them when there is food. The young animals must also have some sort of signal to tell their parents when they are hungry. The social animals which live together in groups need even more signals, and they possess

The fine courtship display of the peacock is often put on for the benefit of humans and other animals, as well as for the peahen.

The gaping yellow beaks of the young blue tits inform the parent that the chicks are hungry and stimulate her to provide food.

a complex "language" of sounds and other signs. Many of these signs are concerned with food and feeding. Others warn of danger, but most of them are probably concerned simply with keeping the group together. These signals may not be very obvious to us and we may even miss them altogether, but they play a very important part in the life of the social creatures.

There are four main groups of signals: visual, chemical, auditory (sound) and tactile (touch). Each kind may be used for any job, in other words a job performed by a visual signal in one group of animals may quite easily be performed by a

chemical signal in another group of animals.

Visual signals

Visual signals are usually the easiest for us to detect and interpret because our eyes are relatively good. Chemical and auditory signals are much harder for us to pick up and understand because, compared with those of other animals, our noses and ears are not at all sensitive.

Although visual courtship displays are most highly developed in birds—for example, the peacock—they are by no means confined to these creatures. Many fine displays are put on by fish, and the courtship of the three-spined stickleback is not difficult to follow in an aquarium tank.

Male spiders are generally very much smaller than their mates, and when they go courting they have to ensure that the females recognize them as mates and not as food. Male wolf spiders identify themselves by an elaborate display of "semaphore", in which the legs or the enlarged palps (sense organs attached to the mouth) are waved about in front of the female.

Living lights

The European glow-worm, which is a beetle and not a worm at all, is one of several nocturnal creatures which attract their mates by means of flashing or glowing lights. The male glow-worm is a winged insect, but the female is wingless. At night the female stations herself in the grass and emits a pale greenish light from the hind end of her body. Any male flying nearby will detect the little spot of light and drop down to the female. But, although the female can switch her light on and off, she does not necessarily switch off when

the male arrives. I have often picked up a glowing female and found a male in attendance, and two or three males can sometimes be found clustered around a glowing female, suggesting perhaps that she does not always find the first arrival to her liking.

Visual defence

Visual signals are also used as a form of defence—for example, the behaviour of the eyed hawkmoth is an instance of defensive display aimed at another species, while,

The eyed hawkmoth is well camouflaged when at rest, but if it is disturbed it reveals the large eye spots on its hind wings and sways backwards and forwards. This display frightens off most of its enemies.

when aimed at other male members of the same species, the claw waving of the fiddler crab is also a defensive display designed to help defend a territory. Gregarious and social animals use visual signals as warnings to the rest of the group. The bobbing white tail of a scurrying rabbit, for example, will immediately cause the other rabbits to take flight. A similar system is used by the pronghorns of the North American plains. When one of them senses danger the white hairs on its rump are erected and it starts to run. The bobbing white patch is clearly visible to the rest of the herd.

The wingless female glow-worm attracts the male by a greenish light produced at the tip of her abdomen.

The male fiddler crab's greatly enlarged claw probably attracts the female, and warns off other males from his territory.

Chemical signals

It has been known for a long time that some animals attract their mates by giving off scent, but recent work has shown that chemical signals have many other roles to play in animal life. Scents and other chemical materials, which are released by an animal and which produces specific reactions in other members of the same species, are called pheromones. Their big advantage over sound signals is that, once produced, they last for a relatively long time, and they also have an advantage over visual signals in that they can turn corners.

Moth collectors frequently use a freshly emerged female as "bait" for males. This technique, known as assembling, works especially well with the European emperor moth and with many of the large silk moths. The female pushes her scent gland out from her hind end and the scent diffuses out into the air. The male moths have remarkably sensitive antennae and they can detect just a few molecules of the scent. When this happens they begin to fly upwind in the direction of the scent, but if they lose it they fly erratically

again. Those moths that pick up the scent again will begin to fly upwind, and sooner or later some of them will find the female. The efficiency of this method was brought home to me recently when I was studying the European emperor moth. I released six marked males about half a mile downwind of a freshly emerged female and then drove back to her in the car—a distance of about three-quarters of a mile by road. One of the marked males was already fluttering around the female when I arrived.

Such chemical attractants are usually very complicated substances and only a few of them have been identified. Each species has its own attractant and will not normally be attracted by that of another species, although closely related species do sometimes pair up.

The majority of butterflies rely on sight to bring them together, but most males have special scent-releasing scales on their wings and the scent, wafted to the female by the fluttering of the male's wings, probably sets in motion the female's mating behaviour.

Scent is also responsible for bringing many male and female mammals together. Dog owners will know

The female emperor moth (top) emits a scent which is picked up by the highly complex antennae of the males.

only too well that a bitch in heat will attract all the dogs in the neighbourhood.

Male mammals very often use scent to mark the limits of their territories. Antelopes, for example, possess various kinds of scent gland which provide these marking scents. A similar use of scent is shown by ants and termites, which lay down scent trails for the other members of the colony to follow. These trail scents may be detected either by the organs of smell or by the organs of taste.

Chemical signals are extremely important among the social insects and among some other gregarious creatures. The "queen substance" produced by the queen honey bee is a pheromone which is licked from her body by her attendant workers and which is then spread throughout the colony by way of the food exchanges which are continually taking place between the insects.

Alarm scent

Many animals, including ants and termites, produce special "fear scents" when they are alarmed. These pheromones are very volatile substances and they spread rapidly through the community, causing the animals to take appropriate action. The scents soon disperse, however, and allow things to return to normal as soon as the danger has passed. Some of the shoaling fishes also release "fear scents" when alarmed or when injured. An injured minnow —one which has been grabbed by a larger fish, for example—gives out a characteristic scent which sends the rest of the shoal scattering in all directions.

Communication by sound

Sound signals are used by many different groups of animals, but this method of communication is most developed among birds. Whole books have been written about bird songs. The cock birds are usually the best musicians, and one of the major functions of the song is to attract a mate. The same song may also serve to defend a territory and to drive away other males. Bird song is therefore at its best during the mating season, and many species stop singing once they have a brood to look after.

Grasshoppers and crickets also use sounds to attract their mates. Only the male can "sing" as a rule, and females can often be attracted by playing a recording of the male's song. As with birds, each species has a distinct song, and it is easier to identify some grasshoppers by their songs than by their looks. Many species have a special courtship song which they use when a female is close at hand. The grasshopper has a row of very small teeth on the inside of his back legs and, when he moves his legs up and down, these teeth strike a hard vein on the wing and produce the sound. You can imitate the action by drawing the teeth of a comb over the edge of your finger nail. Crickets produce their songs by rubbing their wing bases together, but this still involves drawing a row of teeth over a hard edge. This method of sound production is called stridulation.

Warning calls

Sound signals to warn of approaching danger are especially well developed between parents and their offspring. On hearing their parents' alarm calls, many young animals will hurry to their parents or take some other defensive action. Some animals possess several alarm calls, each one being used as a warning

against a different kind of danger. The Californian ground squirrel, for example, utters one kind of chirp if it sees a hawk overhead and a completely different chirp if the enemy is a snake. The squirrel will even indicate the position of the snake by moving towards it, but few other animals are able to inform each other about the direction of an enemy. Dr Niko Tinbergen, one of the foremost authorities on animal behaviour, tells an amusing story of how this fact was demonstrated to him when he was watching herring gulls. He was using a hide which had been standing empty for some time and which had been adopted by the young birds as their hideout. The parent birds spotted Dr Tinbergen in the hide one day and sounded the alarm. The young birds rushed into the hide and crouched there around the very danger of which they had been warned.

Warning signals are also well developed among various social and gregarious creatures. Several species post "look-outs" around the community, and these sentinels warn the rest of approaching danger by a variety of sounds. Even if there are no actual sentinels, the group may be protected by the alert behaviour of all the members. Any individual sensing danger will give the warning signal. The members of many rodent species, for example, warn each

The red spot on the bill of the adult herring gull acts as a releaser for the chicks. They peck at it and thereby stimulate the adult to give them food.

other by drumming on the ground with their hind feet. Beavers slap the water with their broad tails.

Tactile signals

The majority of animals, including man, try to avoid touching most other members of their species. Some animals, however, convey information to each other by touch. The female stickleback is finally persuaded to lay her eggs in the male's nest when he nudges the base of her tail with his snout. The males of various web-spinning spiders use a form of tactile signal to tell the female that they have come to mate with her and not to be eaten. The female normally rushes out to kill anything that enters her web, but the male pulls and plucks the web in such a way that the female "understands" his intentions and allows him to approach.

The sense of touch is especially important among the social animals. Ants communicate a great deal of information to each other by rubbing their antennae together. Some of the information exchanged is undoubtedly in the form of chemical signals, but it is necessary for the insects to touch each other. Honey bees indicate to their hive-mates the direction and distance of a good source of nectar. This is done by "dancing" on the combs, and the other bees pick up much of the information by touching the dancing bees.

If you watch monkeys at the zoo —or in the wild if you are lucky enough to be able to do so—you will often see them sitting quietly and searching through each other's hair. They are not really looking for fleas —in fact, monkeys have very few fleas—and this behaviour is really a way of expressing friendship.

Baboons and most of the other primates often indulge in social grooming. This activity certainly helps to keep down lice and other parasites, but it is mainly a way of showing "friendship" towards each other.

Behaviour of this type is known as social grooming and it is particularly common among the primates, where it plays an important role in keeping the community together. Prairie dogs also groom each other, and various birds preen each other's feathers.

Instinct

When a male emperor moth picks up the female's scent he begins to fly upwind towards the source of the scent. This is a completely automatic reaction, and we must not entertain the idea that the moth is thinking about what it is doing. Thinking is a process confined to man and, perhaps, to some of his closest relatives. The scent received by the emperor moth acts on its nervous system and automatically triggers off the reaction to fly upwind. Automatic reactions of this kind are called instincts. They do not have to be learned because they are inborn and just as much a part of the animal as its shape and colour. Instincts do not show themselves, however, until they are triggered off by the correct stimulus or signal. This signal is known as the releaser because it releases the instinctive act in the animal.

The behaviour of the social animals is complex because they have so much information to give each other and so many kinds of signals. Nevertheless, in those animals which have been studied—only a very small proportion of the total—the "language" has been shown to consist of a large number of releasers, each triggering off a certain element of the behaviour. Social behaviour is therefore controlled largely by instinct although, as we shall see, it is often modified by learning.

Monkeys are generally extremely sociable creatures, and even when not actively playing or grooming they like to sit close to each other.

CHAPTER 2

PARENTAL CARE
AND FAMILY LIFE

Early in the spring the male frogs return to their ponds, and for a few days the air is filled with their noisy croaking. This chorus attracts the females and the animals then pair up in the water. The eggs are soon laid and the partners separate, with little chance of ever meeting again. A similar story can be told about mayflies and some other insects whose males "dance" up and down in dense swarms. Attracted by this movement, and possibly also by scent, a female approaches the swarm and is taken off on a marriage flight by one of the males. As with the frog, the insects separate after mating and have nothing more to do with each other. Relatively few animals form mating swarms like those of the frog or the mayfly, but the instant "divorce" is characteristic of nearly all the so-called lower animals—invertebrates, fishes, amphibians and reptiles.

As well as abandoning their mates, most of the lower animals also abandon their eggs as soon as they are laid. The frog, for example, pays no attention to the spawn once it has been laid and fertilized, and most fishes leave their eggs to float freely in the water. Eggs abandoned in this way are left entirely at the mercy of the environment and many of them perish, either because they find themselves in unsuitable surroundings or because they are easy game for predators. To ensure the survival of the species, however, animals which abandon their eggs usually produce such huge numbers of them that at least a few will survive. The female cod, for example, scatters more than nine million eggs into the sea.

The majority of butterflies and moths lay their eggs on the food plants which they eat. This is an early foreshadowing of parental care because, although there is no actual caring for the young, it does ensure that the caterpillars have the right food from the start. There is, of course, no forethought on the part of the parent. Egg-laying is an instinctive act, often triggered off by the smell or taste of the food plant.

Maternal earwigs

The next stage in the evolution of parental care is represented by those animals which look after their eggs and young for a short time. While the majority of insects abandon their eggs as soon as they are laid, one notable exception is the earwig whose maternal behaviour was noticed more than two hundred years ago by the naturalist Carl Degeer. The common European earwig lays her eggs in an underground chamber and sits over them until they hatch. Experiments performed under artificial conditions

The female midwife toad lays a string of eggs which the male then cares for. He wraps them around his hind legs and carries them about with him until they are ready to hatch.

Hundreds of frogs sometimes gather to spawn in a small pond. Once laid, the spawn is abandoned to the mercy of the environment.

A hunting spider (Palystes) *carries her egg cocoon which she has spun on a leaf.*

have shown that she will even collect up any eggs that have become scattered over the surrounding soil. It is said that the female licks her eggs from time to time and that in fact they will not hatch without this treatment. Perhaps the licking keeps the eggs free from mould spores which might otherwise kill them. After hatching, the young earwigs remain in the nest for several weeks and receive the same careful attention as the eggs. Their mother licks them and it seems probable that she also provides them with some regurgitated food during their first few days. Later on she can be seen taking flower petals and other food materials into the nest. The young earwigs generally leave their nest in late spring, but even then they are not far from their mother's side. She leads them to some suitable feeding ground where they remain as a group until the young ones are almost fully grown.

The parental behaviour of the earwig is far removed from that of the social insects described later, but the social species must clearly have passed through this stage at some time during their evolution.

Paternal fishes

Although most fishes merely scatter their eggs into the water, there are a number of species which take great care of their eggs and offspring. There is therefore a much greater survival rate, so these fishes lay fewer eggs than those which abandon their eggs. Perhaps surprisingly, it is often the male fish which looks after the young. The male seahorse, surely one of the strangest of all fishes, has a brood pouch on his belly and the female lays her eggs there. The pouch closes after the eggs are laid and the eggs get their oxygen supplies from their father's blood circulating in

the walls of the pouch. The "pregnancy" lasts for about five weeks and is followed by the most unusual sight of a male animal giving birth to his offspring. The male pouch opens again and the little seahorses—up to a hundred of them—are forced out, one at a time, by violent muscular contractions. The father usually needs a short rest between each birth.

Among the cichlid fishes of the rivers of Africa and South America, both parents normally help in fanning the eggs and guarding the

Above top: The wolf spider (Pisaura mirabilis) *carries her egg cocoon attached to her jaws.*
Above: Young wolf spiders (Pirata piratica) *ride on their mother's back for their first few days of life.*

A female common European earwig with her newly hatched young.

young. In some species the young fish are even rounded up in the evening and "put to bed" in the nest hollow. Mother normally stays by the nest and signals to her brood by waving her fins. Most of the youngsters return in response to these signals, but father usually has to swim around and pick up the stragglers in his mouth.

The midwife toad

The common toad deposits long chains of eggs in the water and leaves them at the mercy of other creatures, so many of the eggs are eaten. The midwife toad, found in various parts of Europe, avoids much of this wastage by looking after its eggs for the first few weeks. The eggs are laid on land and, after fertilizing them, the male wraps them around his hind legs. There may be up to a hundred eggs in the string. The toads emerge from their burrows at night and the eggs are normally kept moist by the dew, but the male will take them to a pond and give them a dip if the air is too dry. After about three weeks the eggs are ready to hatch and the male again enters the water, allowing the tadpoles to wriggle out of the surrounding jelly and swim away. When they emerge from the eggs the tadpoles are much larger and more advanced than the newly hatched tadpoles of the common toad. As a result, they are better equipped to deal with enemies, and the midwife toad species can survive with a relatively small number of eggs.

Compost heap nests

The various kinds of crocodiles all lay their eggs on or near the river banks. Some species bury their eggs in the ground, while others heap up mounds of rotting vegetation and lay their eggs there. The eggs take as much as four months to hatch and the female crocodile defends her nest throughout this period. When they are about to emerge from their eggs, the young crocodiles signal to their mother by making grunting noises. She immediately begins to dismantle the nest and uncover the eggs. The babies stay very close to their mother for a while and she seems to show them around the territory. As a rule the young ones disperse within a few days, but they have been known to keep together for several months.

Family life

The highest degree of parental care is found among the birds and mammals. These differ from nearly all other animals in that they actually feed their young. The other animals that we have looked at, with the possible exception of the earwig, merely give protection to their young in the earliest stages. Birds and mammals, however, feed and care

for their young over a relatively long period. Both parents are frequently involved, and many of these animals therefore lead a real family life. For the purposes of this chapter, a family can be defined as a group of animals living together and consisting of two parents with their offspring. Any animals which form distinct groups of this kind can be said to lead a family life. We have already seen a simple kind of family life in some of the cichlid fishes, but this is rare and family life is almost entirely confined to the birds and mammals.

The majority of birds are monogamous—that is, they have only one mate during the breeding season—and each pair normally has its own territory which it defends against other birds of the same species. The male usually adopts a territory in the spring and he attracts a female by singing or by some other form of display. The two birds then stay together for the rest of the breeding season and share the work of rearing

the family. There is a good deal of variation, however, in the way in which the work is divided up. In some species the male helps the female to build the nest and he also takes his turn at incubating the eggs. In many of the song birds, however, the male has nothing to do with building the nest or incubating the eggs and he occupies himself solely with collecting food for his wife and family. The female has to join him in this task later on because the rapidly growing young birds need more food than the male can provide by himself.

The young birds are soon ready to leave the nest and to receive their flying lessons. Flying is an instinctive action, but the young birds often need encouragement from their parents. An adult bird can sometimes be seen flying to and fro in front of its youngsters as if to say, "Come on, you can do it". The youngsters soon master the art of flying and they gradually leave their parents. The family thus breaks up,

Left: Some cichlid fishes, known as mouthbrooders, make no nest hollows and the eggs are carried about in the mother's mouth. Even after they have hatched, the young fish return to their mother's mouth if danger threatens.

Right: Family life is often prolonged among the geese and ducks, although the father does not stay permanently with the family. Here, a female merganser is taking her youngsters for a swim.

although the parents may produce another brood before the end of the breeding season.

Life-long marriage

At the end of the breeding season, the pair bonds between parent birds usually break down and, although the birds may continue to associate with one another as part of the non-breeding flock, they no longer co-operate with each other. When the next breeding season comes around the birds will probably find different partners. There are, however, a number of species in which the adults pair for life. Most geese and swans form life-long pairings, and the pairs stay close to each other throughout the year, even when they are travelling in large flocks. Their family life also tends to be more prolonged than in many species because the young birds stay with their parents until long after they can look after themselves. Swan families can be seen swimming together when the cygnets are almost as large as the parents, although they are not mature and still carry their brownish plumage.

Family life among mammals

There is obviously a very strong bond between a female mammal and her young because the female provides her offspring with milk from her own body. There is no possibility of the female's abandoning her young to the care of their father. The latter, in fact, rarely has anything to do with the upbringing of his children, even among the social animals, and many mammals never see their father. The majority of mammals are therefore reared in units which some biologists call "mother families". The duration of these associations varies enormously. Many rodents, for example, leave their mothers almost as soon as they are weaned. Carnivores, such as cats and foxes, stay with their mothers for a much longer time, during which they learn how to fend for themselves and how to hunt their prey. The young animals acquire some of their knowledge by watching their mother, but they also learn a great deal by playing with each other.

The otters of Eurasia and North America form associations which are halfway between the mother family and the true family in which both parents are prominent. The female gives birth to two or three cubs in a grass-lined burrow. The cubs are blind and helpless at first, and it is usually seven weeks before they open their eyes. During this time they are fed and tended by the bitch alone, although the father may visit his mate with food from time to time. He has been known to assist her in digging the burrow, but there are very few records of this occurrence. The young otters leave the

An adult swallow arrives to feed its young. Most birds lead a true family life during the breeding season, with both parents helping to look after the young.

The young mammal's dependence on its mother for milk means that the female mammal must play the major role in rearing the young.

These cygnets will remain with their parents until they are nearly fully grown.

nest for the first time when they are about eight weeks old, but they still keep close together as they explore their surroundings. They follow their mother to the water but, like many human children, they need a good deal of persuasion before they actually take the plunge. While the cubs are still young the father may remain nearby, sharing food with the bitch and sometimes joining her on a fishing expedition. He is not usually allowed near the cubs, however, although there is one report from California that a dog otter took charge of his cubs after their mother had been trapped. The ties between the male and female gradually break down and, sooner or later, the dog otter wanders off. The cubs probably stay with their mother until she mates again, by which time they will be nine or ten months old and quite able to fend for themselves.

A family of gibbons

The majority of the primates, the great order of mammals to which we ourselves belong, are social creatures. They live in communities consisting of several adults and their young, although, as we shall see later, the organization of the community varies considerably. Exceptions to this communal life include the aye-aye and some of the other lemurs, which are solitary creatures, and the gibbons. The latter live in family groups consisting of father and mother with their three or four offspring. Each family has a definite territory which is forbidden to other families. The territory is staked out and defended primarily by vocal signals. A neighbouring family hearing the shouting of the resident gibbons will as a rule keep away, but families do meet occasionally and then they really will fight to defend their homes.

Unlike most primate societies, where the male is the undisputed head of the household, the gibbon parents rule jointly. Both are very protective towards their young and control the activity of the whole family by a series of sounds and gestures. These signals are learned by the young as they grow up. The animals feed mainly on fruit and leaves, together with some insects, and they frequently share their food with each other. They also groom each other and play chasing and other games. These activities all help to cement the bonds between the members of the family. Some American zoologists have recently shown that the animals prefer certain partners when they indulge in grooming or playing, although a given animal might not take the same partner for grooming and for playing.

Young gibbons usually leave their parents when they are six or seven years old, although some leave when they are younger. Some of them pair up with their own brothers or sisters, while others take mates from other families. The new pair might not breed for a year or two, but they still need a territory and so there is a continual shifting of territorial boundaries as new pairs try to fit themselves in. The bond between the

Most young carnivores are playful animals, and it is by playing that they learn how to defend themselves, how to catch food and generally how to survive. These otter cubs will remain with their mother until they are about nine months old.

The lar gibbon lives in south-east Asia and spends most of its time in the treetops. It usually moves about in family groups, each of which has a definite territory.

parents remains very strong throughout their lives and they stay together even after all their children have grown up and left.

Industrious beavers

The beaver colony must surely be one of the most remarkable animal communities because of the way these large rodents appear to collaborate in building their homes and their famous dams. There are two kinds of beaver, one in Europe and one in North America, but they are very closely related and they may be nothing more than geographical races of a single species. The colony almost always consists of a simple family, with father and mother and two litters of offspring. The maximum size of the colony is about fourteen animals.

The female, normally the stronger of the two sexes, often adopts a stretch of river as her territory before she acquires a mate. She marks the boundaries of her chosen home with a strong scent. The male is attracted by the female's chemical "signposts" and, after a certain amount of courtship, the animals pair up. Pairs are usually formed during the summer and, once they are formed, there is a life-long state of harmony between partners. They sit and sleep close to each other and they can often be seen to "kiss" each other. They also communicate by using very high pitched sounds, often beyond the limit of the human ear. A great deal of social grooming takes place, which helps to cement the bonds between the two animals as well as keeping their waterproof coats in good condition.

The parent beavers are, of course, the central pillars of the colony and they are extremely attached to their territory. In the autumn they are affected by the urge to build and, led

The typical beaver lodge is a dome-shaped pile of mud and branches rising up from the stream-bed. The animals live in a large chamber above the water level but all the entrances are below water. Mud is not plastered over the top of the lodge and air can thus reach the living chamber.

by the male, begin to construct themselves a home. Although the beavers share the work of building the lodge, they do not actually work as a team. Each one is acting on its own instinct to pile up branches and mud at the chosen site.

Underwater larders

Beavers living in the northern regions, as most of them do today, experience severe winters during which their ponds and streams are frozen over for considerable periods. As all entrances to the lodge are under water, the beavers are unable to get up on to the land to collect the aspen and willow bark which are their main food. Adult beavers probably live mainly on their own fat deposits during the winter, but young animals are unable to do this and we find that the beavers normally lay in a store of food for the winter. This store is in the form of small branches stacked near the entrance tunnels. The animals therefore have food available, even if the pond is completely frozen over.

The dam builders

Dam-building is undoubtedly the activity for which the beavers are most famous. The dams are built downstream from the lodge site and, by holding back the water, ensure that the lodge entrances are always sufficiently covered. The beaver ponds so created also help other animals, including man. They are valuable refuges for other forms of wildlife and assist in the conservation of water supplies by evening out the flow of water. Beaver ponds can also be useful fire-breaks.

But beavers do not always make dams. Lars Wilsson, a Swedish naturalist, tells how his captive beavers start to build dams when

they hear the sound of running water, and it seems that such sounds are necessary to trigger off dam-building in the wild. If the territory is based on a quiet, slow-moving stream the beavers may not build a dam for some time, but sooner or later some branches will get trapped and the water will start to ripple over them. This is the stimulus the beavers need and they will start to build. Branches, stones and mud are all used in the construction of the dam, but the most striking thing is that the beavers actually go out and fell trees for use in the dam. Young trees are quickly felled as the beaver goes to work with its chisel-like teeth. The trunks are cut into sections and then dragged, rolled or floated down to the site of the dam. As in lodge-building, the beavers work as individuals, although there are reports of two beavers taking turns at felling particularly large trees. Dam-building goes on until the animals can no longer hear the water pouring through or over the structure, and a very solid and efficient dam results.

Everyone helps

Beaver young are normally born in the spring and a litter may contain up to eight of them, although the usual number is about four. The young ones leave the lodge and begin to gnaw bark when they are about six weeks old, but they are still cared for by their parents. The mother spends a great deal of time combing and grooming them because they have not yet learned the art of comb-ing themselves and keeping their fur in good condition. For a further month or two the young beavers explore the territory and then, as autumn approaches, they help their parents and any older offspring there may be to repair the lodge and

to collect the winter's stores. They also help with running repairs to the dam.

Young beavers normally stay with their parents until they are two years old. It was thought at one time that the parents forced them out, but it is now known that the young beavers have an instinctive urge to leave home and find territories for themselves. They pair up during the summer and autumn and produce their first litters when they are about three years old.

Towards social life

There are various ways of defining social life, but it will be adequate if we define the social animals as being those species whose adult members live permanently together in organized groups of more than two. Such a community could obviously develop from family life if the young animals stayed on with their parents or with each other until they themselves started to breed. Alternatively, a social community could develop through the banding together of strangers. Both ways have undoubtedly played a part in the evolution of the social animals.

The European badger, which is found throughout most of Asia as well as Europe, has a life history roughly halfway between the true family life and the completely social life. The adults almost certainly pair for life and they live in a system of burrows known as a sett. These are normally on sloping, well-drained ground and very frequently in wood-land. They may be used for many years and by many generations of badgers, during which time the setts become greatly enlarged and acquire numerous entrances. Some really large setts may accommodate two or three badger families.

The young badgers are born early

in the spring and most females produce twins or triplets. The youngsters start to come above ground during April and at first are closely watched by their mother. They are very playful and they indulge in many rather boisterous games. Where two families share a sett the youngsters often play in one large group and they may, in fact, cause the parents to move in with each other, although each family normally remains a distinct unit at this time. During the summer the young badgers explore farther and farther afield and it is usually these youngsters that dig new burrows around the edges of the setts. The whole family may move from one sett to another and there is a tendency for families to band together for a couple of months during the autumn. The cubs may have started to prepare their own setts by this time and they usually leave their parents for good during October, although a female cub may occasionally take up residence in a large sett with her parents. By the end of October the adult pairs are usually leaving the communal sett and going off to their own homes —not necessarily the same setts as they had the previous year—which they refurnish with fresh bracken and other dried vegetation before settling down for the winter and raising a new family in the spring.

It is very likely that many of the fully social animals passed through the stage exhibited by the badger, full social life having evolved through the development of longer periods of association.

A badger family emerges from its sett at night. In the autumn two or more such families may join forces.

CHAPTER 3

WHY LIVE
IN COMMUNITIES?

Family life is obviously of great advantage to a species because it ensures that the young animals are brought up in safety and get a chance to learn how to protect and feed themselves before they go out into the world. The value of living in larger communities is less obvious, however, and it is pertinent to ask why communal life should have evolved before we go on to look in more detail at some of the communities themselves.

Leaving aside the social insects, whose communities are run on quite different lines from those of other animals, we find that social life is best developed among the mammals. This is undoubtedly associated with the greater brain power of mammals because social life, as already defined, involves some sort of organization and this in turn necessitates some degree of learning ability. Four mammalian orders stand out from the others for their highly organized communities. These are the primates, the carnivores, and the two orders of hoofed mammals, or ungulates. The sizes of these communities vary from three individuals to many thousands, and we use names such as pack, clan, troop and herd to describe them.

Social life is less highly developed among birds and relatively few species have anything resembling the structured communities that we find among the mammals. Outside the breeding season many birds certainly gather together into huge flocks but, although the birds definitely benefit from this arrangement, the flocks are only loosely organized. They have no leaders and there is no co-operation between the members. It is perhaps better to call these birds sociable rather than social, the difference being that sociable merely means fond of company whereas social implies some degree of organization. A few bird species remain sociable during the breeding season as well, but the majority are strictly territorial at this time and each family lives in its own distinct territory, even if, as with many sea-birds, the territories are very close together.

Many amphibians congregate in large numbers at the breeding season and some may hibernate together because they are all drawn to the same site. Reptiles may also hibernate in groups and some of them bask in the sun together, but none of these animals forms a true society such as we find among the birds and mammals. Sociability is, in fact, more noticeable among the fishes than among the amphibians and reptiles. Many fish species swim in large shoals or schools and, although there may be no leader, the animals react to each other in several ways to maintain the group.

Schooling fishes seem to be irresistibly attracted by moving objects of their own size, and it is in this way that large shoals are formed, like this one of monocle bream. The seemingly regular spacing between the fish is brought about by their reactions to the bright patterns of their neighbours and the vibrations set up in the water.

33

Thresher sharks use their long tails to round up the small fishes on which they feed. Two or three sharks sometimes join forces to do this, but such co-operation is very rare among fishes.

The advantages of communal life

Community life must clearly be of some advantage to the animals or else it would not have evolved; and the benefits are of several kinds. It is a well known saying that there is safety in numbers and it is certainly true that group life affords some protection against predators. One commonly cited example concerns the starling and the peregrine. The latter is a bird of prey which swoops down on its victims at high speed. It always selects isolated birds, because diving into a flock would result in serious injury to the peregrine. The starlings take up close formation when one of these predators appears and thereby prevent an attack. Many fishes also prevent attacks by swimming closely together and confusing their enemies with flashing colour patterns and rapid movements. It has been clearly shown, too, that fish eat fewer water fleas when the latter are in dense swarms than when they are scattered.

The musk ox is a shaggy, heavily-built animal that roams in small herds over the tundra of Greenland and North America. The young animals sometimes fall prey to wolves, but when a wolf pack appears the adult members of the herd normally form a living barricade by standing in a circle around their young, facing outwards—a wolf pack has very little chance of getting past such a solid array of bone and horn. The musk oxen also adopt the same tactics to protect their young from blizzards, the long shaggy coats of the adults forming an efficient curtain around the young in the centre.

Living in a group can also help animals to find food because several animals can cover a much greater area in the search much more effectively than a single one can. Those which find good feeding grounds lead or direct their fellows to them, no animal excelling the honey bee in this respect. Predatory animals, such as wolves and cheetahs, benefit from hunting in groups because together they can bring down larger prey than if they hunted singly. The animals can also work in relays and tire out a victim.

The thresher shark is not a social or gregarious creature but is remarkable for the way in which it sometimes joins forces with others of the species to round up herrings and other shoaling fish on which it feeds. Threshers have extraordinarily long tails, up to ten feet long and accounting for half their length, and they use these as whips to shepherd their victims into a tight bunch. Having done this, the sharks then take a bite or two at the bunch, probably also eating up those fish that are stunned by the lashing movement of the tails. Although many fishes live in large shoals, this is one of the very few examples of mutual assistance between the individual members of a species, apart from co-operation in rearing the young.

Division of labour

Animals which live alone have to do all their own work. They have to find their own food, make their own nests, rear their own young, and keep a constant watch for enemies. Group living opens up the possibility of dividing up the work, so that one individual does one job and another performs a different task. This is, of course, what happens in human society and such specialization is very useful. We would not get on nearly so well if we all had to build our own houses, make our own clothes, grow our own food and teach our own children. Among the other animals, division of labour is best developed in the social insects, and we shall see that some ants and termites are structurally adapted for carrying out certain tasks. Division of labour among the birds and mammals is largely concerned with

looking after the young. Wolves and hunting dogs normally leave one or more of their numbers behind to guard the young while the rest of the pack go off to hunt. Penguin chicks huddle together in large groups when their parents go off to feed, but a few adult birds always seem to stay behind. There are similarities between this behaviour and the behaviour of human parents who leave their children in nurseries while they go shopping, but there is not much evidence that the penguin "guards" take much notice of the young. The penguin "crèches" are probably largely self-protective because of their large numbers.

Rabbits and many other grazing animals are said to post sentries around the outside of the group

Top: Soldier termites stand guard as workers lay the foundations for a new tunnel. The large jaws of the soldiers are well suited to their job of defending the colony.

Bottom: Young emperor penguins gather into groups when their parents go off to feed. A few adults remain with the youngsters, although they take little notice of them.

while they are feeding. The sentries are believed to keep a watch for enemies and to warn the rest of the group if any appear. Some animals undoubtedly do take turns keeping watch in this way, but there is another explanation. Any animal at the outside of a feeding group is likely to sense danger before its fellows at the centre, and one would therefore expect those animals around the outside to give the alarm

The members of a minnow shoal stay together partly by scent.

whether they had been given the job of sentry or not.

Stimulating company

Dr J. C. Welty, an American biologist, studied the behaviour of goldfish living singly and in groups, and he discovered some rather interesting facts. The fish living in groups each consumed considerably more water fleas than those fish living alone. The group-dwelling fish also grew more rapidly than the others, but this was not due entirely to their increased food consumption. Another experiment showed that the fish in the groups grew more rapidly even when all the fish had the same amount of food. The presence of other individuals of the same kind obviously stimulates the metabolism of a fish and makes it

more efficient. It has also been shown that goldfish do better in water that has already contained goldfish than they do in fresh water. This shows that the stimulus is a chemical one—in other words, the fish give off a pheromone which stimulates their feeding activity.

Chickens and other animals have also been shown to feed and grow more rapidly when they are living in groups, while nest-building and similar activities may be considerably speeded up by the proximity of other individuals. Gannets and other sea-birds living in the centre of a colony frequently complete their nests and rear their young more rapidly than the birds nesting near the edge of the colony, where there are not so many neighbouring birds. This increase in efficiency as a result of living in a group is called social facilitation and it can also be witnessed among humans, although the mechanisms involved in bringing it about may be rather different. Athletes train harder when they are together than when they train alone, and I am quite sure that my digging rate goes up when other members of the family are in the garden with me.

"Herding instinct"

Some early naturalists believed that the members of a herd or pack of animals were held together by an instinctive desire to be with other members of the same species. This idea of a "herding instinct" was popular for some time but, like so many other early ideas, it had to be abandoned when more detailed investigations into animal behaviour were carried out. Animals are certainly attracted to each other, but it is quite clear now that the bonds between them result from a variety of instinctive reactions and not from a single broad "herding instinct". Scent, sight and sound all play a part in keeping communities together.

Schooling fishes

Probably something like four thousand different kinds of fishes—about one fifth of all the known species—have evolved the schooling or shoaling habit, showing that communal life of this kind must have considerable advantages. The majority of fish shoals are leaderless assemblages without any social structure—in other words fishes are sociable rather than social creatures—but their members are nevertheless held together by very strong instincts. If you watch a shoal you will see that all the individual fish are swimming parallel to and at a fixed distance from each other. When disturbed they will move off or turn in perfect unison, almost as if they were joined together and part of a single being. The main factor involved in keeping the fish together is sight, and it has been shown that fish will not school when their eyes have been covered. The majority of schooling fishes seem to be irresistibly attracted to moving objects of their own size, and it is this instinctive reaction that draws them together into schools, sometimes with more than a million individuals. Attraction to objects of their own size means that young and adult fish normally form separate shoals, and it is rare to find fish of different ages swimming together.

Although sight is the main factor involved in the schooling of fish, other senses play some part. Scent, for example, is known to be involved in keeping minnow shoals together. Vibrations are also important, and it has been suggested that the precise spacing of fish within a shoal is maintained by reactions to the vibrations of neighbouring fish. The fish have, after all, a very efficient system for picking up vibrations in the water. This is the lateral line system, which is in the form of a slender canal running along each side of the body.

Knowing one's friends

The members of a fish shoal are held together purely by instinctive reactions which require no learning. Such instinctive responses can also be found among birds and mammals but a good deal of the social behaviour of these animals has to be learned, especially among the mammals. As we shall see in the next chapter, most social mammals live in relatively small units and they are not automatically drawn to all other members of the same species. Young mammals grow up in the group and come to accept the other members as an integral part of their surroundings. Helped perhaps by the pleasures of mutual grooming, the animals learn to recognize each other as "friends" and they tend to stay together after that.

Territory and home range

As well as forming attachments to each other, many social animals form attachments to certain areas. This is another mechanism which helps to keep the members of a group together. Two distinct kinds of "land holding" can be distinguished. A territory is an area which

Many birds form large flocks, like this one of Arctic terns, but they rarely have any leaders and there is seldom any co-operation between the members.

is inhabited by a family or group of animals and which is defended against other animals of the same species. Fierce fights sometimes break out on the boundaries of the territories, but fighting is rather wasteful as far as the species is concerned so the majority of animals have evolved some form of threatening behaviour which is sufficient to drive away any trespasser. The threatening behaviour may take the form of a warning scream, a harsh stare or a show of strength.

The second kind of land tenure found among animals is the home range. It is frequently larger than a territory and the animals roam all over it, but they do not defend it. The home ranges of neighbouring groups very often overlap and the groups often meet each other.

The "peck order"

The members of an animal community usually form a definite hierarchy, with the stronger animals at the top and the weaker ones at the bottom. The stronger ones "bully" those below them on the social scale and often chase them away from food. Such a hierarchy was first described by the Norwegian biologist Schjelderup-Ebbe in 1922. Working with a flock of domestic hens, he noticed that there was one hen which pecked all the others without getting more than an occasional peck back in return. At the other end of the scale there was a poor hen which received pecks from all its associates and led a very miserable life indeed. The rest of the hens occupied intermediate positions, each one pecking those below it but not, or only rarely, attempting to peck those above it in the hierarchy. This kind of social structure was called a peck order, but today it is more usual to call it a rank order because it is found in many animals and not just in those which can peck.

A certain amount of fighting and squabbling goes on in a community before the rank order is established, but even then the fighting is rarely serious. A show of strength or of self-assurance is often enough to decide the position of two hens.

A straightforward hierarchy or rank order, such as that described above, is found mainly in those communities in which the animals still act largely as individuals. Swordtail fish, for example, develop what has been called a "nip order". The fish at the top of the hierarchy is a tyrant and he always gets the best food, while the fish at the bottom is harassed by all the others and often starves to death. In this kind of hierarchy, however, more often

Fish shoals are generally leaderless associations, but their members are held together by chemical and visual signals, such as the bold patterns of these snapper fish.

than not the tyrant is overthrown.

The social mammals normally have a more stable hierarchy than the other animals, and the dominant animal is not a tyrant but more of a central figure around which the rest of the community revolves. The dominant member of a group of mammals is normally an adult male and he holds his position because he can put on a greater show of strength or aggression than his fellows. They soon learn that he is the master and they normally allow him first choice of food and of resting places. In many communities the dominant male also has first choice of mates, and he may completely monopolize the females. He is often aggressive towards other adult or sub-adult males but, unlike the tyrant hen, he does not attack those animals at the bottom of the scale. In fact, many dominant males allow the youngsters to play all over them.

The dominant male may also assume responsibility for protecting his group from outside enemies, and we often find members of the group gathering round him. Such "loyalty" is an important factor in keeping the community together. Fights occasionally break out between members of the group and then the dominant animal turns peacemaker, threatening to intervene himself if the fighting continues. The dominant male therefore plays a major part in maintaining harmony within the group, but he is not necessarily the leader—that is, he does not perforce decide the direction of the group's movement, nor does he necessarily walk in front of them. The leader of a group has to be particularly sensitive to danger and must also have some kind of responsibility towards the other members. Adult females, especially those with young, often fulfil these conditions better than other animals,

and we find that many grazing animals are led by females even if there are dominant males among them. Young animals growing up in a group soon learn to recognize the dominant animal and the leader and they also find their own place in the hierarchy as they grow up.

The existence of a rank order is of value to the species in several ways. Once the order is established there is little squabbling within the community because each member knows its place and does not attempt to cross a higher ranking animal. The animals therefore live peacefully, and it has been shown that some species eat more and grow more rapidly when living in a stable hierarchy.

Although rank orders are usually very stable once they are established, there are certain circumstances in which an animal can change its position in the hierarchy. When an animal gets old or sick, for example, it may lose its self-assurance and it will soon drop down the social ladder. On the other hand, Konrad Lorenz showed that a female in his jackdaw colony immediately rose in rank when she became "engaged" to a male of higher position. In fact, she took on the same rank as her intended mate. Pregnant mammals and females with young also tend to rise in social rank.

Jackdaws live in small colonies and have a definite rank order, although the dominant male is aggressive only to his nearest rivals. He shows considerable "concern" for his flock and will even go out and search for a missing member.

CHAPTER 4

SOME MAMMALIAN COMMUNITIES

The aim of this chapter is to describe the social organization of a number of mammals, chosen to show the wide variation found among this class of animals. Many other examples could have been described, and although the examples chosen here represent various levels of social organization, it is not suggested that they form any kind of evolutionary sequence—except perhaps among the seals. Social behaviour, just like any other feature of an animal, has evolved to suit the habitat in which the animal lives and it does not indicate the animal's position on the evolutionary tree of life. Social behaviour is often closely linked with reproduction and may appear only during the breeding season.

The rabbit

The European rabbit has been studied in great detail by the British naturalists H. N. Southern and R. M. Lockley. The typical rabbit community is based on the warren— a system of burrows usually situated on sloping ground and rarely far from trees or other reasonable cover. The size of the warren and the number of animals in the community vary considerably. The individual animals do not have a great deal of contact and, apart from the members of a pair, they are rarely found within a foot of each other. But in spite of this rather individual way of life, the rabbits maintain a stable and well-ordered society, based mainly on the territorial behaviour of the females, or does. Each doe adopts a small area of the warren as her home and, if necessary, digs herself a burrow. Once she has settled in she rarely goes more than a few yards from her home and she defends the immediate surroundings of her burrow against other females.

The bucks of the colony engage in fierce fighting during the autumn and a dominant animal soon emerges. He becomes the absolute ruler of the warren, or of a part of it if it is a large one. The other bucks are tolerated, however, as long as they show proper respect for the "king". This respect involves moving away when he approaches, and he will attack any other buck which comes within about a yard of him. The rank order breaks down in mid-summer when the breeding season is over and the animals are beginning to moult, and the animals seem not to notice each other for two or three months. Then, in the autumn, the fighting starts again and a new rank order is established, with many new males and often with a new ruler.

The dominant male has first choice of living quarters and he

The red stag defends his harem with a deep-throated bellow which keeps all but the strongest rivals away. The large antlers come into play if the stag is challenged by a rival, the two animals putting their antlers together and engaging in a trial of strength.

41

usually takes up residence in the central area of his territory. The doe living there becomes his mate and takes on the role of the dominant doe, although rank order is far less obvious among does than among bucks. The subordinate males are allowed to make their homes within the dominant male's territory as long as they do not come too near to his own burrow. They take up residence with the other females in the colony, most of the subordinate males having only one female. The ruler and one or two of the other males may associate with more than one female, but even then each male seems always to have a favourite mate. The pairs keep together, more because of their attachment to their adopted territory than because of their attachment to each other.

Each evening the "king" patrols his territory and the other bucks normally move out of his way. But sometimes fights do occur and occasionally the dominant buck is beaten. His conqueror then takes his place and moves into the central residence, taking over the "queen" doe as well. The death of a dominant doe, however, does not lead to such an upheaval. A young female, possibly one of the "queen's" daughters, moves in to take her place with the king, and the other adult females remain in their chosen territories, making no attempt to move into a better one.

The rabbit young are born mainly in the spring and early summer, and their upbringing is entirely the responsibility of the females. Although blind and helpless at first, the young rabbits grow very quickly and mature in about four months. The adults then frequently chase them away from the warren, especially the young males. The young females often stay around the edges of the warren and dig burrows there. The warren thus gradually increases in area. Young males may move to other warrens and be accepted there, but they must reach an age of at least eighteen months before they have any chance of achieving a dominant position.

Underground townships

The prairie dog is another burrowing creature, but it is friendlier to its fellows than the rabbit. Whereas physical contact is minimal between rabbits, prairie dogs seem to need such contact and they are forever "kissing" each other.

Once found all over the plains of North America, the number of prairie dogs is much smaller now because they are regarded as pests by farmers and ranchers and have been exterminated over large areas. They are not dogs at all but ground-living squirrels, and they get their

common name from the dog-like bark which they use as a warning call. There are five species, of which the best known is the black-tailed prairie dog, recently studied in great detail by John King in South Dakota and Ronald Smith in Kansas. All five species have similar habits but they live in different regions. The white-tailed prairie dog, for example, tends to live at higher altitudes than the black-tailed species.

Their burrows form a network of tunnels, between three and four feet below ground, and the entrances are normally surrounded by conical mounds which protect the tunnels from flood water. These mounds are not just accumulations of soil dug out from the tunnels, for they are rapidly rebuilt when they are damaged. Visual contact is also important in the prairie dog's life and the vegetation around the burrows is always kept down to a level which enables the animals to see their neighbours—and also to spot coyotes and other predators.

The burrows are grouped into colonies known as towns, each of which may cover many acres—and in former times some of them covered many square miles—but each town is always distinctly separated from neighbouring towns. There is usually at least a mile between them, unless there is a river or some other barrier, and it is very rare to find a prairie dog outside his own town. Each town is divided into several wards whose boundaries normally follow minor geographical features. A slight slope, for example, or a string of bushes, may divide one ward from the next. Animals do sometimes cross from one ward to another but this is unusual. Each ward is further divided into coteries and these are the real social units of the population.

For the most part, each coterie is

A group of prairie dogs. They are not true dogs, but ground-living squirrels.

less than an acre and contains four or five adult animals. There is usually only one adult male and he is the dominant animal, but there is no obvious rank order. The coterie territory is vigorously defended, especially by the dominant male who patrols the area each day. He is normally the first to emerge from the burrow in the morning and the last to go to bed at night.

The animals living in each coterie are all very friendly towards each other, except in the breeding season, and they are always visiting their neighbours' burrows and helping dig new tunnels or make new mounds. When they meet, the animals "kiss" each other by baring their teeth and putting their mouths together, thereby identifying themselves as friends. Social grooming is very common, the animals nibbling and pawing each other for long periods. Any animal may groom any other member of the coterie and the young actually "ask" to be groomed by nuzzling up to the adults.

Prairie dogs are quite noisy creatures and communicate by means of a wide range of calls. There are several kinds of warning bark which warn all the ward members and not just the members of one coterie. The females also have a territorial call which they use when they meet a "foreigner" in the coterie. This call summons the adult male who soon banishes the invader by using threatening behaviour. Visual contact is also important, for the sight of other prairie dogs seems to give the animals a sense of security.

The breeding season occurs in February and March. The females become antagonistic towards each other while pregnant or nursing their young, but once the young are above ground friendship returns and the animals mix freely. The young very often end up in the wrong burrows at night, but they are well looked after by any female in the coterie and they come to no harm.

Harems on the sea shore

Seals are marine mammals, but they have not completely lost their ties with the land because they all come ashore to breed. A few species, such as the leopard seal and the Ross seal, are solitary creatures but the majority are gregarious. During the breeding season, when they form dense colonies on the shore, many of the species exhibit a very definite social structure.

The grey seal, which lives around the coasts of the North Atlantic, forms loosely-organized harems in the breeding season. The males arrive at the breeding grounds from

A group of grey seals.

August onwards and establish small territories. The females, many of them already pregnant, arrive a month or so later and they are gathered into harems by the territory-holding males. The pups are born shortly afterwards and the males then mate with some or all of the females in their territory. The territories are not vigorously defended, however, and it is usual for a succession of males to rule over each harem during the breeding season. The females may also move from one harem to another, and they "belong" to the male of whatever territory they happen to be in.

The elephant seals, so called because of the trunk-like snout of the male, have a much more stable harem system than the grey seal. There are two species: the southern elephant seal of the Antarctic, and the northern elephant seal which was once almost extinct but which is now spreading along the coasts of California and neighbouring regions. The two species look much alike and their behaviour is basically very similar. The following account

is based mainly on George Bartholomew's studies of the northern elephant seal.

Outside the breeding season, the northern elephant seal forms dense colonies containing males and females of all ages. They usually contain a fair number of sea lions as well. There is no sign of a dominant animal or of a hierarchy, but there is very little squabbling and the general atmosphere is peaceful. Many of the animals are playful and they chase each other in and out of the water. When they are resting on land they like to be in close contact with each other, and one can

Fur seals form vast breeding colonies, often with thousands of individuals. The colony is broken up into numerous harems, with one adult male defending up to fifty of the smaller females.

Male fur seals patrol the boundaries of their harems and defend them by roaring loudly and making threatening gestures.

Male elephant seals in combat position.

Female elephant seals with their pups.

often see young ones sleeping on the backs of the elders. Even the sea lions sleep on top of the sea elephants.

The animals move to their breeding grounds in January and establish breeding colonies which, compared with the peaceful non-breeding flocks, are scenes of tremendous activity. The pregnant females usually arrive first, followed by the young unmated females and the adult males. Yearling and juvenile animals are not normally found at the breeding grounds, although small numbers may occur around the edges. The males roar and bellow at each other, using the "trunk" to produce a far-carrying sound. These vocal challenges, accompanied by a series of threatening gestures, drive the weaker males to the edge of the breeding colony and leave a few of the largest males

as beach-masters spaced out more or less regularly among the dense mass of females. Each dominant male has an average of about thirteen females around him and he continues to defend his position by roaring and gesturing. Intruders are chased off, but the beach-masters do not constantly patrol their territories and herd their females. They seem more concerned with maintaining their positions among the females than with keeping a definite group together.

The pups are born soon after the breeding colonies are established and are cared for by their mothers for several weeks. The beach-master protects his females and young against intruders, but he is rather careless as he lumbers around and pups are often crushed as he crawls over them. About a month after the pups are born the females are ready to mate again. All the females, including those mating for the first time, come into heat together—probably as a result of some pheromone or chemical stimulus. The beach-masters mate with all the females around them, but the females are not necessarily faithful to the male in whose harem they have been living. Many females on the edge of the colony will mate with the less dominant males banished to the periphery by the beach-masters. After mating the breeding colonies gradually break up and the animals return to the sea. The pups stay with their mothers for a few more weeks, but then all parental ties are broken.

Fur seals and sea lions are more mobile on land than the true seals and elephant seals because they can use their hind limbs to lope around. The beach-masters use this extra mobility to patrol their harems and keep their females together. Unlike the elephant seal, it is not enough for the male fur seal simply to be among the females: he must rule over them.

The grazing herds

We use the word "herd" in connection with several kinds of animal community, but we use it most frequently in connection with the

Common seals have no social structure even during the breeding season. Mating takes place at random and there is no dominant male.

47

of female company. They enter the home ranges of the female herds and each stag tries to round up a harem of hinds. There is a great deal of competition between the stags and the air is filled with their deep-throated bellows. The roar of a fully grown stag is enough to keep most rivals away, but another fully grown stag will not be put off by a roar alone and the two animals will engage in a "pushing contest". This trial of strength is a sort of tug-of-war in reverse: the two rivals put their antlers together and push. The weaker animal soon gives way and departs, leaving the winner to round up the hinds. This is a never-ending business because the hinds have no loyalty to the stags and they are always wandering away. Despite the stag's presence, the leadership of the herd remains with the original female leader. She warns of approaching danger and the rest of the herd will follow her when necessary.

At the end of the rut the stags leave the home ranges of the females and go off to re-form the male herds. The hinds remain on their ranges and the calves are born early in the following summer. The calves stay with their mothers for about eighteen months. Then the young males leave to join the male herds, while the females stay with their original herds throughout their lives, taking an active part in the rut when they are about two and a half years old.

Joint rulers on the plains

The wildebeest, or brindled gnu, is one of the commonest of the African game animals. Herds containing many thousands of animals roam the plains of central and east Africa and, although probably out-numbered by Thomson's gazelle, they must produce a greater weight,

hoofed mammals or ungulates—the deer, antelopes, horses and so on. Some of these animals can be found in herds containing thousands of individuals, but such vast assemblages are usually only temporary. The true social units are much smaller and there are many different kinds.

The red deer, or wapiti as it is called in America, is basically a forest-dwelling animal, but it is equally at home on moors and mountains. For most of the year the adult males and females live apart. The males live in nomadic leader-less herds, while the females remain on their home ranges and each small herd possesses a definite leader.

The breeding season, or rut, occurs in the autumn when the stags, now with fully grown antlers, leave their all-male herds in search or biomass, than any of the other game animals. The information presented here is based mainly on the researches of Lee and Martha Talbot, who made a detailed study of the animals in the Serengeti region of Tanzania.

Outside the breeding season, the animals roam the plains in large or small herds, according to the abundance of food. These herds are quite unorganized and their leaders are continually changing. It seems that the animals keep together partly by means of a tarry odour given off by glands on the feet and partly by their instinctive desire to follow other large moving objects.

Very large herds build up at the beginning of the dry season and the animals move westwards on their annual migration in search of fresh pastures. The rutting season starts at about the same time and, during pauses in the migration, some of the adult males acquire breeding herds. A single male may round up about fifty females and young, but more often the males work in twos and threes and herd as many as 150 females and young into a tight bunch. There is no hierarchy among these co-operating males, and no fighting. They are joint rulers and they share all the females. Each breeding herd and its immediate surroundings constitute a territory and the males defend it by circling round the females, their heads held high. The resident males often gang up to chase away a challenger. Females straying from the herd will be shepherded back by one of the males as long as they do not stray too far: a male will rarely enter another territory to bring back a wayward female, and she for her part will be quite happy to join a neighbouring group.

Some males are able to maintain their breeding herds while the animals are on the move, but the breeding herds normally break up when the migration gets under way again and the females all merge into one large moving mass. The next time they stop, the rounding-up process starts again and a new set of breeding herds is established for a few days, usually with a different set of adult males. Only a few males are actively engaged in herding activities at any one time. The rest of them, including all the young males, group together to form male herds a little way from the breeding herds. Males rarely mate until they are three years old because they are not large enough or strong enough to round up and maintain a breeding herd before that age.

At the end of the rutting season the males lose the urge to form breeding herds and the whole population merges into the structureless herds already described. These herds gradually move back to the east for the rainy season.

Zebras and wild horses

It used to be thought that the vast herds of zebra found on the African plains were unorganized and leaderless communities, but recent work, notably that of Hans Klingel, has shown that the zebras are really

One of the commonest of the African game animals, the wildebeest forms herds sometimes containing many thousands of animals. They are often associated with other game animals such as the zebra and elephant.

Although hundreds of zebras can often be seen together on the plains of Africa, the true social unit rarely contains more than a dozen animals.

very highly organized animals. The great herds that come together from time to time are actually composed of numerous distinct social units.

Two kinds of social unit are found among the zebras: the breeding group, often called the "family group", and the stallion herd. The breeding group consists of an adult stallion and up to six mares and their young. Some of these groups contain more than a dozen members, but six or seven is more usual. The stallion is the dominant animal but, as with most of the other hoofed mammals, the group is led by a female. There is a well marked hierarchy among the mares and the dominant mare takes the lead when the group moves off. The other mares follow in accordance with their rank and the stallion usually brings up the rear. These breeding groups are very stable, the bonds between the individuals being maintained by a good deal of mutual grooming and by greeting ceremonies whenever

they meet after a short separation. The animals are also constantly calling to each other while they are grazing, thus ensuring that they do not get too far away from each other. At least one animal seems to remain alert when the group is resting and it gives out a series of warning calls if it spots a lion or other predator.

A female zebra remains with her breeding group throughout her adult life, even when she is old and weak. There are many reports of compassionate behaviour in these animals, the herd members helping to support a sick colleague and even going in search of missing members. Such behaviour does not extend to the stallion, however. When the stallion becomes old or sick he is rapidly replaced by a younger one and he goes off to join one of the stallion herds which normally contain up to fifteen males. These herds are always led by an adult but there is no apparent hierarchy within them.

The only truly wild horse alive today is Przewalski's horse, a stockily-built creature which retains a rather precarious hold on life on the steppes of western Mongolia. Like the zebra, it lives in small herds but the stallion is the leader of the herd. The other so-called wild horses, including the various breeds of pony found in the British Isles, are all descended from domestic stock which went wild. Many of them are still very much under the influence of man.

Some of these semi-wild horses, such as the mustangs of North America, live in herds which contain males and females of all ages. The herd is led and dominated by an adult stallion who banishes the rest of the stallions to the edges. There is also a rank order among the

Some wild horses, such as these New Forest ponies, live in small herds, each with a single stallion. Some stallions are removed each year, and many herds have no adult males.

A herd of African elephants — distinguishable from the Indian species by their larger ears.

The semi-wild mustangs of North America live in herds, dominated by an adult stallion.

mares, although in larger herds it is often complicated by many triangular relationships. Other semi-wild horses also live in herds, each with a single stallion.

Purring elephants

Elephants live in herds containing anywhere from ten to fifty indivi-duals, but it is quite common to find smaller groups consisting of a mother and her offspring of various ages. The members of a herd are usually related to each other in some way. The smallest herds may, as already mentioned, consist only of mother and offspring, while the largest herds build up when these offspring stay together after producing young themselves. A few outsiders may also join the herd from time to time. Herds often exchange members when they meet. Occasionally one also comes across much larger groups, but these are merely aggregations of small herds which have joined forces temporarily in an area of good food supply.

The herd is always dominated and led by an adult female with young, the bull elephants playing little part in the social organization. In fact, the older bulls often take themselves off and live apart from the rest of the herd for much of the time. Some bulls become quite solitary, especially in old age. Elephants are entirely vegetarian and they eat vast quantities of leaves and grass. While moving about in the bush they keep in contact with each other by deep rumbling noises, rather like purring, which seem to come from the stomach. This "purring" stops when an elephant is disturbed and the silence alerts the other members of the herd—an interesting reversal of the usual use of sound for warning purposes.

Elephants are very reluctant to leave their sick or injured behind, and there are many reports of them helping each other in distress. The most frequent examples of this compassionate behaviour involve attempts by two or more animals to lift an injured colleague and help it along by supporting it between their own bodies. Surprisingly enough, this kind of behaviour has hardly ever been recorded in the females, although these are the ones which show the strongest social bonds in other respects. The females are especially co-operative in the business of rearing their young.

Cats—solitary and social

The leopard and the tiger, in common with almost all the other cat species, are solitary creatures. The adults come together only for a short period in the breeding season. The only truly social cat is the lion. Lions live in small groups, called prides, which contain up to twenty animals and sometimes more. Usually there is only one adult male, but in a large pride there may be two or three. Each pride has a territory, the size of which varies according to the abundance of zebra and wildebeest—the lion's major food. Where food is plentiful the territories are small, but they may reach fifty square miles in other places. The male lions defend their territories very fiercely against strange males, the fights often ending in the death of one of the animals. There are also numerous fights within a pride, for the lions have not yet learned to live with each other in complete harmony.

A male lion may mate with all the lionesses in his pride, but the normal behaviour is for a male to pair up with one female at a time. The pair keep together for several weeks or months, often until the cubs are born, but then the male finds another mate. There is a much stronger and more permanent bond between the females in a pride. Many of them are closely related to each other and they help each other to rear the cubs, even suckling the cubs of another lioness. The male has nothing to do with rearing the cubs and he very rarely does anything about keeping the pride in food. Nearly all the hunting is done by the lionesses, who often join forces to stalk and bring down their prey. But the male, by virtue of his position as ruler of

Although often called the king of the jungle, the lion lives mainly on the grassland and the savannah.

The cheetah, recognizable by the black stripe running from the eye to the mouth, is sometimes found alone, but the animals usually live in small groups.

Lions pair up for several weeks, but the male does not help after the cubs are born and, in fact, rarely does any hunting at all.

the pride, is the first to dine from the carcass.

The cheetah, long believed to be a solitary animal, has recently been shown to have a loosely organized social life. It is something of a link between the solitary leopard and the truly social lion. Cheetahs are sometimes found alone, but they usually move about in groups of three or four individuals. A group may consist of a mother and her young, or of adults—probably a grown-up family. The adult groups always contain at least one male.

A group of cheetahs does not defend a territory, but the animals ensure that they keep away from the other cheetah groups by leaving scent signals in the form of urine. Cheetahs coming upon the smell of fresh urine—indicating the nearby presence of another group—will automatically move away.

The members of the group are all on very friendly terms with each other, the "friendship" being maintained by a good deal of licking and social grooming. There is no rank order, although one of the males is the accepted leader and the group will not move off until he does. Another member of the group will sometimes move a short way away and try to get the others to follow, but he will give up and rejoin the group if the leader shows no interest.

Young cheetahs stay behind while their mother goes hunting and they do not start learning the technique until they are about six months old. They are nearly a year old before they start to contribute to the family larder. At the age of eighteen

months they usually leave their mother. They may leave singly, but more often they go off together and form a new group.

Hunting packs

African hunting dogs are found all over the savannah lands south of the Sahara. The animals live in packs of up to sixty individuals, although most packs number about twenty, and they feed on any game they can get. Gazelles are their main food but they will also pull down wildebeest and zebra. Each pack has a large home range, often covering hundreds of square miles, and somewhere within the range there is a favourite breeding area to which the animals return when any of the females is about to have a litter. The ranges overlap, but the packs normally keep apart because they can smell each other.

It has often been stated that there is no rank order within a pack of hunting dogs, but recent prolonged observations by Hugo van Lawick and Wolfdietrich Kühme have shown that a rank order does exist. As Hugo van Lawick pointed out, the members of a pack are very well known to each other and it is rarely necessary for one dog to assert its dominance over another. Prolonged watching is therefore necessary before one can uncover the rank order. Submissive gestures by the lower-ranking animals include bowing the head, wagging the tail and rolling over on the back—all of which could easily be interpreted as friendly greetings. The problem is further complicated by the fact that both dogs often go through the same ritual when they meet and one cannot pick out the higher-ranking individual. Such behaviour might well mean "I will help you", rather than "I accept your superiority". If

Hunting dogs live in packs of up to sixty individuals, and they hunt over a wide area. Elaborate ceremonies take place before the hunt, and these help to unite the animals and strengthen their co-operative behaviour.

Timber wolves, found in most northern regions of the world, live in small packs of up to a dozen animals. The pack is often a family unit, but additional adults are frequently present. There is a well-defined rank order with an adult male as the dominant animal. The wolves feed mainly on rodents but they often combine to pull down large animals such as caribou and moose.

so, it is clearly in the interests of the pack as a whole.

The females have a separate rank order and it is generally much more obvious than that of the males. High-ranking females often bite their subordinates, and Hugo van Lawick found a good deal of jealousy over the young. When a low-ranking female had a litter the dominant female even tried to take over the pups for a while.

Hunting dogs usually go off to hunt in the evening and again in the early morning, just before sunrise. Before setting off the dogs go through their elaborate greeting ceremonies which, like the howling ceremonies of the wolves, serve to unite the pack for the hunt. The dominant male nearly always takes the lead as the pack moves off and he normally decides the direction in which the pack goes, but any member of the pack may be at the front during the later stages of the hunt. The dogs usually move quite slowly until they near their quarry, and then they break into a run. The first run often stops short of the prey if the latter are in a large herd, but it serves to set the prey scampering away and the dogs are then able to pick out an animal which is weaker than the rest. They concentrate on this animal in the second run and

usually bring it down fairly quickly. At other times the hunting dogs will split up and chase several animals, but each dog keeps an eye on his colleagues and will abandon his own chase if he sees that a colleague is having more success. In this way the dogs all gradually converge on the weakest of their prey.

Young cubs do not accompany the adults on a hunt. They stay around the breeding dens with their mothers and sometimes with other adults as well. When the hunters return the young ones run up to them and "beg" by dancing and wagging their tails, at the same time rubbing their snouts against the lips of the returning hunters. The latter then turn aside and bring up large chunks of meat which are greedily seized by the youngsters. The females who have been on guard at the dens also beg for food, although they do not get as much as the youngsters: the hunters seem to prefer to feed the young dogs. Adult males rarely have any success with their begging, however, and Kühme found that the males who had been left at the breeding dens went out and followed the trail of the returning hunters until they found the remains of the kill. They did not touch other carcasses on the way.

Howling monkeys

Most of the primates are intensely social creatures, but they exhibit a wide variation in the structure of their communities. At one end of the scale we find the gibbon family, already described in Chapter 2, and at the other end we find baboon colonies with more than a hundred individuals. Social groups containing just one adult male and a number of adult females and their young are commonly called bands. Larger groups, in which there are several adult males, are usually called troops or clans.

The howler monkeys live in the tropical forests of South and Central America. They live in small clans, and their social structure was first studied by Dr C. R. Carpenter on Barro Colorado Island—a most important research centre which was created during the construction of the Panama Canal and made into a nature reserve in 1923.

Howlers, as one would expect from their name, are very noisy animals and they communicate with each other by means of a wide variety of sounds. A loud roar from the adult males warns the others of approaching danger, while a rapid grunt attracts the rest of the group and draws their attention to something—perhaps a particularly good crop of figs, of which the howlers are very fond. A more or less constant clicking sound given out by all the animals helps to keep the clan together when moving about the tree tops. The animals howl very loudly at dawn, as if laying claim to a particular area for the day, for they are very intolerant of contact with other clans. If another clan is heard nearby the members will close up together and set up a deafening chorus of howls and roars. This behaviour normally keeps the clans apart and so, although the howlers do not defend a particular area of forest, they do defend their immediate surroundings wherever they happen to be.

Within the clan the adult males are the dominant animals and they control the movement and activity of the others. There is no obvious hierarchy, however, and the males take turns leading. The females normally mate with all the males in the clan and, as the animals breed throughout the year, there is usually at least one female with a young baby. Young babies are of great interest to all the females and they gather round the new mother whenever they stop to rest. A howler rarely leaves the clan into which it is born and strangers are rarely accepted. Clans do split up, however, if their numbers rise too much, and sub-adult males sometimes leave to wander alone for a while. They may rejoin the clan later or they may tag on to other clans and follow them around until they become accepted as members.

Baboons

Baboons are large powerful monkeys living in various parts of Africa. There are several species and two very distinct kinds of social

Hyenas live in clans of up to a hundred animals, each clan being ruled by a female. The clan has a definite territory and the boundaries are marked with scent. Contrary to popular opinion, hyenas are not cowardly creatures that feed only on carrion. A clan of spotted hyenas will chase and kill a wildebeest.

structure. The savannah, or chacma, baboon lives in the southern and eastern parts of Africa, inhabiting fairly open bush country and frequently damaging cultivated land. Its social life has been investigated by several people, notably by two American biologists, Washburn and Devore. The animals live in large troops, consisting usually of between fifty and eighty individuals but sometimes numbering more than a hundred. Each troop has a home range of up to six square miles and the ranges of neighbouring troops usually overlap. The troops often meet at a water hole or at a particularly good feeding ground, and they show a good deal of interest in each other, but they do not normally exchange animals. Each troop is a closed community.

At first the baboon troop might seem a rather disorganized group, mainly because of the large number of animals, but it is really very well organized. There is a definite hierarchy among the males and very little fighting ever takes place between the adults. The dominant

The savannah baboons live in large troops and devote much time to social grooming. The animals are rarely more than a few feet from each other.

males are usually the largest and strongest members of the group, although aggressiveness is just as important in acquiring rank. The dominant males get the best food and have first choice of resting places. They also get most attention in the way of social grooming; a male has only to sit down and there will be a group of females clamouring to groom him. Social grooming is very important to all baboons and they devote several hours a day to this activity. They appear to derive great pleasure from it and it obviously plays a big part in holding the members of the troop together.

When on the move, the leading animals are always the smaller adult males, and these are followed by females with juveniles or children. Behind these come the females with babies and the dominant males, followed by more females and another group of subordinate males which brings up the rear. Subordinate males may also be found around the edges of the troop. The females and the dominant males are therefore protected on all sides by a ring of adult males. When the baboons are attacked by lions or leopards they usually head for the trees, but under these circumstances the females and their young normally move off first. This leaves the adult males to bring up the rear, between the females and the enemy.

Although there is a definite hierarchy among the male baboons, the dominant males do not monopolize the females. A female in heat will mate with any of the males in the troop, especially during the first few days, but she eventually pairs off with one of the dominant males and they move off towards the edge of the troop for a few days. Most of the young baboons are fathered by the dominant males.

The hamadryas, or sacred baboon,
favours dry rocky places and lives in Ethiopia, neighbouring parts of north-east Africa and also on the other side of the Red Sea, in southern Arabia. The social unit of the hamadryas is much smaller than that of the savannah baboon, consisting of just one adult male together with about four females and their young. The unit is therefore a band or harem, although it is usual to find several bands in close proximity, giving the appearance of a much larger troop. This is especially so at night, when more than a hundred animals may gather to sleep on the ledges of a small cliff. The male dominates the members of his band in every way, although he puts up with a good deal of boisterous play by his children. The females normally follow the male very closely, but they sometimes wander off and then are "punished" by a bite on the neck, which is the male's way of showing them that he is the boss. The dominant males are quite tolerant of each other's presence and there seems to be no rank order among them when the bands come together in this way. There is a great deal of social activity when the animals wake up early in the morning, and the members of different bands mix freely. Young animals play with and groom members of other bands, but adults confine such social behaviour to members of their own band. The animals gradually disperse during the morning, each band going off to find food for itself, but they normally come back to their favourite sleeping quarters each night.

Man's closest relative

The chimpanzees are probably our nearest living relatives and they are certainly the most intelligent of the apes. An adult chimp has the

Chimpanzees live in the rain forests of Africa and spend most of their time in the trees.

intelligence of a child of five, and until the age of four a chimp is actually ahead of a human child of the same age. Chimps have long been favourites in zoos, but we knew very little about their natural lives until Jane Goodall, a British naturalist, went to live among them in Tanzania and got to know many of them individually.

Chimpanzees live in rather loosely knit societies containing up to about eighty individuals, and such a clan may have a home range of between six and fifty square miles, according to the habitat, the populations being densest in the forests. The clan has no definite leader and is continually breaking up into sub-clans which wander alone for a while and then join together again.

Although there is no obvious leader in the chimp society, there is a well-marked rank order, with the adult males nearly always dominating the females. The highest-ranking male is usually the one who puts on the greatest show of strength and makes the most noise with his hooting and screaming but, like man, the animals are not above using artificial aids to boost their egos. Jane Goodall tells how a low-ranking male transformed himself into the clan ruler by banging oil cans together and thus making more noise than the other males. Fights break out every now and then if a subordinate animal offends a superior by stealing food or by failing to get out of the way, but the fights are usually short-lived and the subordinate animal is sent away with a friendly pat on the head. Friendly relationships are the basis of chimpanzee society and a great deal of time is spent in social grooming. Physical contact and the reassurance that it brings are of great importance to the animals, especially to the youngsters.

Chimps are completely promiscuous animals, and a female in heat will usually mate with all the males in her clan as well as with any outsiders she meets. The rank order of the males has no effect on their mating. One particular male may stay close to the female for a while after her mating urge has subsided, but he gradually loses interest and goes back to join the rest of the males. Bringing up the children is not for the males and they are just as likely to go off to another clan if they can find another receptive female.

Gentle giants

Early accounts of the gorilla suggested it was ferocious, but more recent observations have shown it in a new light. The studies of the hunter Fred Merfield and George Schaller, an American biologist, among others, have shown it to be a peaceful animal, attacking only when provoked and usually more interested in getting away from the intruding human.

The gorillas live in small clans, usually with between six and seventeen members. These clans are very stable and wander over home ranges between ten and fifteen square miles in area. Each clan is ruled by an "old man"—an adult male at least ten years old. Males of this age always have a certain amount of silvery-grey hair on the back and they are therefore often referred to as "silverbacks". The larger clans may contain more than one silverback, in which case the oldest one is usually the leader, even if he is not the largest. The typical clan also contains one or more blackbacked males. These are between six and ten years old and, although they are fully adult, they are not mature enough to rule the clan. The clan is completed by a number of adult females and their young of various ages. There is a well-marked hierarchy and there is hardly any fighting within the clan. A subordinate will almost always move out of the way of a superior animal and give up a favourite resting place. Such actions are usually entirely spontaneous, but the superior animal sometimes has to resort to a tap on the shoulder before the other one moves out of the way. Silverbacks are always dominant to blackbacks and females, but there is no fixed relationship between the blackbacks and the adult females. An old blackback will dominate a younger female, while an old female will be superior to a young blackback. The females also have a rank order among themselves, although it is not particularly obvious. It is based partly on age and temperament, but the possession of a young baby puts a female right to the top. She is then superior even to an old blackback. All adult gorillas are superior to the juveniles in the clan, and the juveniles have their own rank order based on age and size.

The dominant silverback makes all the decisions in the clan—when to get up in the morning, which direction to take, when to eat, and where to rest. Some of the subordinate males occasionally go off by themselves and they may try to get the rest of the clan to follow, but there is almost total obedience to the clan ruler who indicates his intentions by various movements and attitudes and an assortment of grunts. They communicate by voice when moving about in the forest, and the males advertise their position by beating their chests—an action which draws the other gorillas together. When the members of the clan are close together, however, communication is almost entirely by gesture. For example, when the male gets up and walks quickly in a certain direction he is instructing the others to follow him. The blackbacks normally bring up the rear when the clan is on the move, but any animal can be the leader in an emergency.

Although the "old man" dominates all the other members of his clan, this dominance does not extend much further than having the right to the best resting places. He has no great claim to the females, which are free to mate with any of the males in the clan.

The stability of the clan is based on numerous individual relationships between the members, and there are several reports of gorillas assisting injured colleagues and helping them to keep up with the clan. The "old man" is a good parent and guardian, defending his subjects to the end if

necessary, although man is the only real enemy of the gorilla. Infant gorillas, less than three years old, actually seek out the dominant silverback and romp around him. He fusses over them and plays good-naturedly as long as they are not too boisterous, but if they overstep the mark he will send them away, merely by altering his expression and staring harshly at them.

There is a deep relationship between the females and their young, lasting for up to six years. The females cuddle their babies a great deal, just like human mothers do, and they also spend a lot of time grooming their offspring. It is during this long period of "childhood" that the young gorillas learn the meaning of all the sounds and gestures used by the group. A deep relationship also exists between the adult females in the clan and they spend a lot of time in close contact with each other. Social grooming, however, is uncommon among adult gorillas, only odd minutes here and there being devoted to this activity.

The female gorilla normally gives birth about once every four years and so the clan grows relatively slowly, especially when we take into account the fact that the males often leave the clan. Large clans do split up from time to time, with half following one silverback and half following another. New clans also come into existence when a lone male lures females away from established clans. Very old silverbacks are usually challenged by younger ones and when they are defeated they leave the clan to end their days as solitary animals. A clan which loses its only silverback frequently joins forces with another clan or else it is taken over by a mature lone male. The females are hardly ever found without the company of a mature silverback.

64

CHAPTER 5

SOCIABLE BIRDS

No one who has seen the sky blackened by flocks of starlings or seen trees break under the sheer weight of roosting budgerigars can doubt that birds like each other's company. Nevertheless, most bird flocks are structureless associations: members come and go more or less as they please and the flocks have no constant composition. There is no co-operation between the members and little or no recognition of one individual by another. Most birds are thus described as sociable. Very few species show anything like the true social or communal life that we have seen among the mammals.

The majority of birds are monogamous, meaning that they have only one mate at a time. They spend the breeding season in family units. Most of them are strictly territorial at this time, each family inhabiting a certain area and defending it against other members of the same species. This territorial behaviour is associated with the need to collect sufficient food for the young birds without having to go too far away. Some species remain territorial throughout the year, but once the breeding season is over most birds abandon their territories and mix freely with each other. Many of them, especially those which have mixed and unspecialized diets, band together in large feeding and roosting flocks. Some birds, such as the rook, remain sociable throughout the year, feeding together during the breeding season and defending only the immediate surroundings of their nests. No bird is known to be sociable during the breeding season and solitary or territorial at other times, although some of the sea-birds certainly form denser colonies during the breeding season than at other times.

Gannets by the thousand

Like all sea-birds, the gannet has to come ashore to breed. It normally chooses isolated islands for this purpose and vast numbers of birds congregate in one small area. The colony consists of thousands of small territories, each inhabited and defended by a pair of birds which have their nest at the centre. Although the nests are only three feet apart, the birds do not co-operate at all. In fact, they do just the reverse, for they are always ready to steal nesting material from unwary neighbours. Nevertheless, there must be some advantages in nesting so close to each other because it is not lack of space that causes the birds to congregate so densely. A dense colony may exist on one part of an island while the rest of it, although apparently similar, is quite devoid of gannets.

Social stimulation is probably the

Nests of the common weaver bird hang in dense colonies with dozens of nests in a single tree.

the wrong territory could have painful consequences.

The males arrive at the breeding grounds first and begin to establish territories. Old males normally go back to the territories they held the previous year, while the younger ones which are breeding for the first time seek out abandoned territories or else form new ones on the edge of the colony. The females arrive somewhat later and again the older ones seek out their previous territories and their old mates. Young females fly to and fro over the colony searching for lone males—either young ones or "widowers" whose mates have not returned.

The territories are no more than a yard square but they are just large enough for the birds to build their nests and rear their young without interference. Some fighting goes on while the territories are being established, but once ownership is settled the occupiers are left to mate and rear their young in peace. The nests are little more than piles of seaweed and other assorted materials cemented together with the birds' droppings. They are very regularly spaced, so that each sitting bird is just out of pecking range of its neighbours.

Gannets nest in dense colonies on isolated cliffs (top). Each pair defends a small territory just large enough to hold the nest and the sitting bird (below).

biggest advantage derived from colonial nesting. Apart from those birds nesting at the edge of the colony, the gannets all lay their eggs at the same time and these all hatch at the same time. This synchronization may well ensure that the young birds appear just when the food supplies are at their height. Colonial nesting also helps the gannets to deal with the egg-stealing gulls which are their main enemies. It is a brave gull that attempts to land in the centre of a gannetry. Even the gannets themselves have to be careful how they approach their nests, for landing in

The Adélie penguin

The Adélie penguin is the commonest and most widespread of the seventeen penguin species. It lives in Antarctica and the surrounding islands and breeds in vast rookeries containing thousands of pairs. The species has long been known to Antarctic explorers but it has only recently been studied in detail by the scientist William Sladen.

Adélies mate for life, but the pairs stay together only during the breeding season and by a remarkable feat of memory and navigation find each other again in the snowy wastes. The birds arrive at the breeding grounds in September and October and it is usually the experienced males—the ones which have bred before—that return first. They find old nest sites even if they are buried in the snow and they usually remain faithful to these sites, although there is a tendency for the birds to move in towards the centre of the rookery if there are vacant sites. Inexperienced birds—four-year-old ones which have not bred before—usually take up nest sites near the edge of the colony. When the females arrive the males put on a big display of flipper-waving and they squawk noisily. The older females pick out their mates quite easily and join them at the nest site. Males often "flirt" with strange females but all flirting stops when the "wife" arrives and the strangers depart. Only if his mate is very late will the male seriously entertain a strange female.

Both sexes collect stones for the nest and they are not averse to stealing them from their neighbours. Two eggs are laid and then, with the male taking the first incubation period, the female goes off to sea to feed. She may be away for about eighteen days and there is much excitement when she returns to take over the eggs and allow the male to break his fast, for by this time he has not fed for five or six weeks. He returns to the nest after a couple of weeks and the female goes off again. The eggs hatch while she is away and she soon returns with food for the chicks. The parents then take turns bringing food every couple of days or so. When they are about a month old the chicks leave their nests and they all herd together in large groups, or crèches. This helps to keep them warm and it also deters predatory birds such as skuas, but there are no real guardians:

no adults are appointed to look after the young. There may well be some unsuccessful breeders still trying to establish a nest site, but these birds pay no attention to the chicks in the crèches. The parent birds are often away together and they feed their chicks less often at this stage, but they can still recognize their own chicks and they will not feed any others. A chick whose parents have died is itself doomed to death from starvation.

When the chicks are about eight weeks old the crèches are beginning

Guillemots, in common with gannets and most other sea-birds, nest in dense colonies on the cliffs.

to break up and by the ninth week the chicks are ready to enter the water. The rookeries then disintegrate, about three months after they assembled, and the birds spend the next nine months fishing around the pack ice.

Tree-top colonies

The rook is a large black member of the crow family, distinguished from the carrion crow by its greyish white face and its greyish beak. It feeds on the ground all day, but nests and roosts in the tree tops. Dense woodlands are avoided and the most favoured sites are small clumps of trees or narrow strips of woodland bordering streams and farmland. The birds are intensely sociable and they always nest in colonies. A rookery may contain as few as a dozen nests, but most rookeries are much larger and some contain more than a thousand nests, with a dozen or more in a single tree. Each rookery has its own feeding grounds, usually close at hand, and the birds feed there together. It is very rare to see a rook all by itself.

The nesting season lasts from March to July and after that the rookery is more or less abandoned for a while. The birds still use their normal feeding grounds but they prefer to roost in nearby trees. In September, or even earlier in the more northerly regions, the rooks begin to use their winter roosts, which may be as much as ten miles from the rookery. These winter roosts are used year after year and they draw birds from many rookeries. There may be 15,000 or more birds in the roost each night. They begin to arrive before sunset, which means mid-afternoon in December, and one can see flocks flying in from all directions. Occasionally the birds make straight for the trees but they more often assemble in the nearby fields and move to the trees when it is nearly dark. This is a very noisy business and one always has good warning that one is approaching a roost. The birds take about half an hour to settle themselves down for the night and then all is quiet for a few hours.

The birds begin to leave the roost again at first light and each flock makes its way back to its own feeding grounds. The dispersal of the roost, however, is far less spectacular than its assembly because the birds do not all fly off at once. On their way back to their feeding grounds, many of the rooks pay a quick visit to their rookeries and they may even do a bit of repair work on their nests. They very often visit the rookeries in the afternoons as well, before setting off for the roosts. This attachment to the rookeries is an important factor in the maintenance of the community, and it is responsible for the permanent nature of the rookeries themselves.

As winter begins to turn into spring the visits to the rookeries become more frequent and more prolonged, especially the evening visits. Nest building and repair begin in earnest. The birds are already paired up because they normally mate for life and the pairs stay together throughout the year. Young birds probably pair up for the first time during their second autumn. Both members of the pair collect twigs for the nest but the female normally does most of the construction work. By the middle of February nest building is in full swing and the birds are spending all night and much of the day at the rookery, defending their nests against marauding neighbours who are always ready to move in and steal twigs.

Although each pair of rooks

defends its nest and immediate surroundings, the community which we saw remain together throughout the winter roosting period does not break down. The birds continue to feed together. There are even reports that they destroy new nests if these are built too far from the centre of the rookery, thus forcing the builders—usually young birds—to move back into the main mass of the community. The rook is one of the very few birds to maintain such a close community throughout the year.

The starling

The starling breeds in holes in trees, cliffs or buildings. Although the nests are spaced out as widely as possible, the starling is not really a territorial bird. Even at the height of the breeding season it remains sociable and

large numbers can be seen feeding together in the fields and gardens. Like most birds they will often fight over a tasty morsel of food, but there is no other sign of aggression or territorial behaviour when the birds are away from their nests. After the breeding season, some time in June, the birds become even more sociable and they stay together all day and all night. During the day they usually move about in small flocks, perhaps with only a dozen birds in each, but towards evening the flocks join forces and fly to a communal roost. Some starlings fly as much as twenty miles to their nightly roost, so a large roost may draw birds from an area of twelve hundred square miles and contain upwards of 50,000 birds. The roosting sites are very variable, ranging from reed beds to tall trees and cliffs. Many starlings also roost on buildings, but this is a relatively recent habit which they have acquired during the last fifty or sixty years.

Lie-abed sparrows

Like the rook, the house sparrow remains sociable throughout the year, but it rarely forms large flocks. It is rare to see more than about fifty of them together, but just as rare to see only one or two. The birds usually nest in, or very close to, buildings, but the nests are not evenly distributed: a group of houses at one end of a road may support several nests

Left: Rooks usually choose deciduous trees for their nests, although in exposed coastal regions they will nest in evergreens. Each tree usually supports several nests.

Right: Arriving at the roost, the rooks often circle round for a while before settling. Numbers are sometimes so great that they form a black "cloud".

Many birds, such as these swallows, form migrating flocks. Prior to migration, the birds gradually come together and cluster on trees, telephone wires and so on.

Sparrows get on well together and there are rarely any territorial disputes.

each, while similar houses at the other end of the road may have no sparrows at all. The house sparrow thus forms distinct colonies. The areas chosen are always very close to adequate feeding grounds and, apart from defending the immediate vicinity of its nest, a sparrow has no need to establish a territory. Away from the nests the birds mix freely with other members of their colony and, although they often fight over a large piece of crust, relationships are very friendly.

Two or three broods are raised each year between April and July. At the end of the breeding season the birds tend to go off to the grain fields. They never venture far from cover, however, and the flocks are continually flitting back into the hedges and trees as one or other of the birds gives the alarm. While they are out in the fields the birds often join with flocks from other breeding communities and they set up temporary roosts in the surrounding trees. Later in the year, however, each flock returns to its own breeding area and roosts there throughout the winter. The roost may be in a tree or hedge but in the more northerly parts of their range the sparrows tend to roost in holes. Many of them roost on their old nests and they can be seen bringing feathers and pieces of hay to the nests throughout the autumn and winter. The nests thus reach considerable size, one in the roof of my house being at least as big as a pillow.

Although the shrill twittering of the sparrows is a good alarm clock in the summer—and much more pleasing on the ear than the mechanical kind—the birds cannot be relied on during the winter. Roosting sparrows bed down early in the evening and get up late—a habit which, it has been jokingly suggested, they might have acquired through their association with man!

The most destructive bird in the world

This is how many people describe the red-billed quelea, often simply called the quelea, which is one of the weaver birds and belongs to the same family as the house sparrow. Weaver birds get their name because many of the species make intricate nests by weaving grasses and other materials together. Most of the weaver birds are gregarious and many form large colonies, but none rivals the quelea in the size and density of its flocks. This sparrow-sized bird inhabits the savannah and grassland of Africa, ranging in a narrow band from Senegal to the Sudan and then southwards to cover almost all of eastern and southern Africa. The quelea eats virtually

Male black grouse (blackcocks) congregate on an area called the "lek", where they put on communal displays to attract females. The more males there are, the better the display and the more likely it is to attract the females.

nothing but seeds, and with densities in some places of more than a million birds to the acre, it poses a threat to agriculture almost as great as a plague of locusts. The following information is drawn mainly from the researches of John H. Crook, who has made a detailed study of the quelea in various parts of Africa.

Outside the breeding season the feeding flocks of the quelea are usually composed of about five hundred birds. As with most other birds, there is no social structure, but the individuals are attracted by the sight of each other and also by their constant chattering. These factors help to keep the flock together, but the individual birds rarely make physical contact with each other and usually keep a few inches apart, just like shoaling fishes.

At night a number of neighbouring flocks join up to roost in the trees or tall grasses. Most roosts probably contain just a few thousand birds but some of the roosts in the Nile Valley are known to contain millions of birds every night. When food is plentiful these huge concentrations of birds remain in a relatively small area, but they normally disperse during the day and the smaller flocks cover a very wide area, flying as much as thirty miles from the roost in their search for food. It is unlikely that they all return to the same roost each night.

During the breeding season the birds congregate in the damper equatorial regions, some of the flocks flying several hundred miles to reach the breeding grounds. The breeding birds settle in dense colonies, often covering several square miles, and each male adopts a territory of a foot or two in diameter in one of the trees. As with most territorial animals, there is a certain amount of friction between the birds as they try to stake their claims, but the aggressive behaviour gradually subsides and the birds soon learn to respect their neighbours' boundaries. Once the disputes are over the males settle down to festoon the trees with their neatly woven nests. Several hundred nests may be built in one small tree and the sight of these densely packed nests draws

the other birds into the centre of the colony. The birds will even abandon nests which they have already started in order to move to a denser part of the colony. So great is the weight of the nests in some parts of the colony that the trees are completely broken.

When his nest is almost complete the male quelea is ready to "advertise" for a mate, and he does this by holding his wings out and waving them in what is known as the butterfly display. The same display is used later as a greeting when one of the birds returns to the nest. When a mate has been obtained the two birds complete the nest and the female lays her three eggs. Both sexes help with the incubation. Intruders are threatened if they get

The social weavers co-operate to build a "thatched roof" over part of a tree and they then construct their own nests in the thatch (above). Each nest has its own entrance (below).

too close to the nest but away from it the birds remain extremely sociable and they continue to feed in vast flocks just as they do outside the breeding season.

Bird villages

The quelea nests are very close together, but each nest is a distinct structure and it is made by just one pair of birds. Several kinds of weaver birds, however, construct nests which are actually in contact with each other and which may share a common wall. These birds really have reached the stage at which we can say they lead social lives. One of the most famous of these birds is the social weaver. More than a hundred birds may co-operate to build a thatched roof over part of a tree, and then each pair builds a flask-shaped nest under this roof. More than three hundred nests have been counted in such colonies.

Communal nests

Some of the weaver birds are polygamous, with each male building several nests and looking after several females. Add this to the fact that some of the males build "semi-detached" nests with common walls and you can see that such behaviour is not very far removed from true communal breeding in which several females lay in one nest. Communal nesting is found with some of the babblers, thrush-like birds living in Africa, southern Asia and Australia. They inhabit the scrub and woodland areas and, being poor fliers in general, they spend most of their time on the ground. Several of the species move about in small flocks and they keep in contact with each other by way of their babbling voices.

Writing about the Australian white-browed babbler, D. Goodwin describes how one bird really seemed to become alarmed when it discovered that it was alone. He also describes how the birds preen each other by peering closely between the parted feathers and picking out the parasites. Reports of communal nesting come from several places and show that several species build communal nests. Two or more pairs combine to build a nest and the females then all lay in the one nest. One of the most remarkable of the babbler nests is that constructed by one of the scimitar babblers of New Guinea. It is a long tubular arrangement suspended from a vine and, according to the local inhabitants, the birds all sleep in it as well as rearing their young in it.

Social cuckoos

It is a well known fact that many species of cuckoo lay their eggs in the nests of other birds, but this is by no means the only strange habit found in this unusual family. The ani, a long-tailed black bird living in tropical America, forms true communities very similar to those formed by some mammals. The bird has been studied by D. E. Davis in Cuba and by Alexander Skutch on Barro Colorado.

Weaver birds get their name from the elaborate way in which they weave grass and other material to make their nests. Most of them nest in colonies.

The ani is a noisy bird and it lives in flocks of about a dozen individuals. Each flock has its own territory, amounting to about ten acres, which it defends against other flocks. The birds feed mainly on the ground, eating insects and other small animals as well as fruits and seeds. At night they roost in tight little clumps in the trees. There is no apparent peck order and the birds maintain their friendship by mutual preening. They also seem to show concern if one of their number is injured. The members of a flock combine to build an untidy nest in the fork of a tree. A pair of birds may sometimes keep to themselves, but the ani is usually promiscuous and all the adult males mate with all the adult females. The latter then lay their eggs in the communal nest and all the adults assist in incubation, the males taking the night shift and the females sitting by day. Davis recorded twenty-nine eggs in one nest, and clutches of more than twenty are quite common, but the birds are not able to incubate such large clutches efficiently and many of the eggs fail to hatch. The chicks that do develop are fed by any of the adults in the colony until they are about a month old. Some birds stay in their home territory and they even help their parents to rear later broods.

The ostrich harem

Outside the breeding season the ostriches move around in flocks of perhaps fifty birds, led by a male or a female who decides when and where to feed. The flocks are very often associated with herds of zebra and it is possible that the two species help each other by giving warning of danger. The ostrich's height and eyesight, together with the zebra's sense of smell, create a very efficient early warning system against predators such as lions.

At the beginning of the breeding season the males move away from the feeding grounds and establish territories. Each male then rounds up a harem of up to five females and

scrapes a nest hollow in the ground. After mating, each female lays up to eight eggs in the one nest, producing a clutch of about forty eggs, although most nests contain no more than twenty. Even this number presents some difficulty in incubation and, as with the ani, many of the eggs fail to hatch. The male does most of the incubating, but one or more of the females remain on hand and relieve him for short periods during the day. The eggs hatch about forty days after they are laid and the chicks stay with their father, sometimes accompanied by their mothers as well, for several months. The young birds then band together to form juvenile flocks and the adults go back to the communal feeding grounds and rejoin the feeding flocks. The breeding behaviour of the ostrich is thus very similar to that of many mammals, except that the female mammal plays the major role in looking after her offspring.

The rhea is the South American equivalent of the ostrich. It is considerably smaller than the ostrich, but it has evolved remarkably similar looks and habits. Outside the breeding season the birds run in flocks of up to forty individuals and, like the ostriches, they often mix with large herbivores. At breeding time each male rounds up a harem of about eight females and encourages them to lay their eggs in a communal nest. The harems are less stable than those of the ostrich, however, for the females are promiscuous and they go off to mate with other males. Incubation and care of the young are left entirely to the male who may have to deal with thirty or more offspring, not all of them his own.

The largest living bird in the world — the male ostrich sometimes reaches a height of eight feet. It lives in large flocks on the dry grasslands of Africa.

CHAPTER 6

WASPS AND BEES

A great many insects are gregarious. Butterflies congregate on flowers because they are all attracted by the nectar, while ladybirds often come together in their thousands to hibernate in some dry corner. Aphids cluster around a stem during the summer, not because they are attracted to each other but because they are born there and rarely move far from their place of birth. Dense as some of these aggregations may be, there is no social activity at all. There is no co-operation between the animals and they probably remain more or less unaware of each other.

The mating swarms of various gnats and mayflies are slightly more complex aggregations because the insects are attracted to each other, but there is still no real co-operation between them. The vast swarms of locusts which devastate crops in the warmer parts of the world contain millions of insects but, although there is some degree of interaction between the individuals, the locusts are far from being social insects. They lack the basic requirements, which are care of the young and co-operation between adults.

The most destructive of the various species of locust is the desert locust of Africa and southern Asia. For much of the time the insects are solitary creatures, just like most other insects, but under certain climatic conditions the adults become concentrated into relatively small areas. With each female laying up to 200 eggs, the area is soon densely populated with young locusts, or hoppers. Solitary hoppers are green or brown, but when they are crowded together in a small area they develop bold black and yellow patterns. They also become more active when they are crowded and begin to move about in what are commonly known as hopper bands. They have an instinctive urge to keep together at this stage, perhaps triggered off by the bold patterns, and are said to be in the gregarious phase.

After about a month—longer if the weather is cool—the hoppers change into adult insects, pinkish at first and then becoming yellow. This is in contrast to the sandy colour of the solitary adults, which also have shorter wings. The gregarious adults still retain the urge to keep together and, being winged, they can cover large distances. The swarm settles after a while and the insects eat every scrap of greenery that they can find. The females then lay their eggs and a new hopper band appears, many times larger than the original. The process then starts again and the locust population builds up to plague proportions. The build-up may go on for several years, but unfavourable weather eventually brings it to an end and the survivors revert to the

Wasps of the genus Belonogaster *at work on their nest. Eggs can be seen in the incomplete cells on the right.*

Far right: Many butterflies, such as these small tortoiseshells, congregate on flower heads, but they are in no way social creatures and take no notice of each other.

A young desert locust. Locusts often form dense swarms but they are not true social creatures, and there is no co-operation between them.

solitary phase in the next generation.

Large families

One might be tempted to think that true social life among the insects has evolved from aggregations such as those just described, but all the evidence suggests otherwise. All the known insect societies consist of one or more parents together with their offspring. Insect communities are therefore essentially family groups, although father is usually missing. They have evolved through an extension of the simple parental care that we have already seen demonstrated by the earwig.

True social life, in which the adults co-operate for the good of the whole community, is found in only four groups of insects. These are the wasps, the bees, the ants and the termites. Some of these creatures live in societies of thousands or even millions of individuals, usually the children of just one female, and the societies are so different from the mammalian societies which we have already considered that many biologists liken the community to a single animal and compare each

individual to a cell in a body. This idea, known as the superorganism theory, was first put forward by W. M. Wheeler in the 1920s. It is supported by the fact that the members of the colony are often assigned to different jobs, as are the various cells in the body, and also by the fact that, when separated from the main body of the community, individual members normally die.

Queens and workers

The majority of solitary insects die before their offspring are mature—in

Section through the nest of a solitary bee (Allodape) *in a dry reed stalk, showing larvae in various stages of development.*

the other social insects are all females. In addition to the three main castes—queens, males and workers—the termites and some of the ants have a fourth caste, known as soldiers, whose job is to defend the colony.

Food-sharing regulates activity

The social insects are forever regurgitating food and sharing it with others. This food-sharing, known as trophallaxis, plays a very important part in regulating the activity of the community and keeping the individuals together. By sharing food in this way, all the colony members acquire a particular odour which enables them to recognize each other and thereby distinguish between nest-mates and aliens. The queen's influence is also felt by all members of the honey bee community through the process of trophallaxis. The queen gives off certain pheromones which are licked from her body by the attendant workers and then passed on to the other members of the community. These pheromones have several functions, among them the prevention of egg-laying by the workers. Similar pheromones occur among the ants and termites.

Solitary wasps

The wasps, bees and ants all belong to the order of insects known as the Hymenoptera. There are thousands of different kinds of wasps, belonging to many different families, but they all feed their young on some kind of animal. Most of the wasps are solitary creatures, and they are often known as digger wasps because many of them make small burrows in which they lay their eggs. The sand wasps of the genus *Ammophila* are well known solitary species,

Eumenes, the potter wasp, builds a little flask of clay and stocks it with paralysed caterpillars. An egg is laid on each flask and the grub that emerges feeds on the caterpillars.

many species, even before their eggs hatch. The evolution of social life therefore required a considerable increase in the length of life of the parent insect. Another essential feature of social life is the division of labour within the community. Egg-laying in the insect community is confined to relatively few females, and often to just one individual who is known as the queen. Relatively few fertile males are required in such a set-up, and the majority of insects in a colony are sterile individuals known as workers. Termite workers are of both sexes, but the workers of although their long slender bodies give them an appearance quite unlike that of the typical wasp or yellow-jacket. As their name suggests, they live in sandy regions. The mated female digs a burrow in the sand and then, after closing the entrance with a small stone, flies off in search of a suitable caterpillar for food for the grubs. Each species of *Ammophila* has its own preferred type of caterpillar, which it paralyses with its sting and then drags or carries back to the burrow. Although the sand wasp usually makes a short reconnaissance flight before leaving her burrow, she does not always remember its exact position when she returns and often has to search for it. The great French naturalist Henri Fabre studied digger wasps at great length, and he recorded how one returning female, unable to locate her burrow right away, laid down her burden and started a thorough search. Periodically she rushed back to the caterpillar as if to make sure it was still there and then continued her search. Other entomologists, however, working with other *Ammophila* species, have witnessed the female carrying her burden throughout her search. When the wasp finally discovers her burrow again she opens it, drags the caterpillar in, and then lays an egg. With some species of *Ammophila* that is the end of the story: the burrow is sealed and the wasp flies off to start another. Other species of *Ammophila*, however, practise what is known as progressive provisioning: instead of stocking the burrow with one large caterpillar, they bring smaller ones at intervals as the grub feeds. There is thus some contact between the mother and her offspring—a necessary step in the evolution of social behaviour.

The evolution of the social wasps

The social wasps all belong to the family Vespidae, characterized by the deeply notched eyes and the way in which the wings are folded lengthways when the insects are at rest. We have no direct evidence as to how the social wasps evolved from solitary ancestors, but by studying the

Left: Jean Henri Fabre devoted much of his life to observing the solitary wasps and other insects, and although he did not always make correct deductions from his observations, he contributed much to our knowledge of wasp behaviour. His statue stands in the Provençal village of Sèrignan where he spent much of his life.

Right: A queen wasp of the genus Vespula.

A section of a wasps' nest, showing the tiers of combs and the entrance at the bottom.

Wasps build their nests with "wasp paper" which they make by chewing wood and mixing it with saliva. The picture below shows part of the outer covering of a Vespula nest. Each band of colour is produced by one wasp and one load of woodpulp.

various levels of social organization found in this family today, we can get a good idea of the stages through which the social wasps must have passed.

Many of the vespid wasps still retain their solitary habits, some of the best known being the potter wasps which make neat flask-shaped nests of mud or clay. The nests each contain one egg and they are mass-provisioned with caterpillars before they are closed. Several nests may be built close together and even in contact with each other, but the insects emerging from them do not associate other than for the purposes of mating. As with all the solitary species, males and females are produced in more or less equal numbers.

Wasps of the genus *Stenogaster*, which inhabit the tropical parts of the Old World, make nests containing several cells. These nests are made, like nearly all the social wasps' nests, from "wasp paper", which the insects produce by chewing wood fragments and then spreading out the pulp with their jaws. The nests are made only by the female who then lays an egg in each cell. She feeds the larvae progressively but she does not abandon the nest when the grubs have pupated. She stays nearby and often lays another

batch of eggs in the cells when the first new wasps have emerged. In some species of *Stenogaster*, these new wasps fly off right away, but in other species both males and females stay around the nest for a few days. They do not help their mother with her new brood, but this behaviour is a definite move towards social life. The new wasps then mate and the females, which are fully fertile, fly away to make their own nests.

The African genus *Belonogaster* shows the beginnings of a division of labour and it is a truly social insect. The nest consists of a single sheet, or comb, with perhaps several hundred cells. It is founded by one or more females. They lay their eggs in the cells and feed the grubs with chewed caterpillars—the solitary wasps almost always provide complete caterpillars for their grubs. The new females which emerge are fully developed and structurally no different from their mothers, but during their early adult life they act as workers, collecting food and looking after their younger brothers and sisters. Only in later life do they start to lay eggs. Many of them stay and lay their eggs in the parental nest, but others fly off in groups to found new nests. The males do not usually leave the nests for long and they do no work.

Belonogaster is a rather primitive member of a group known as the polybiine wasps. These are found mainly in tropical America and many of them have a much more rigid division of labour than we saw in *Belonogaster*. There are usually numerous egg-laying females in the nest, but the majority of the wasps are workers. These do not look any different from the fertile females except that they are slightly smaller in the abdomen, but their ovaries are poorly developed and they cannot normally lay eggs. Polybiine

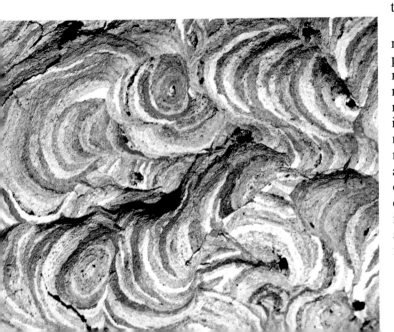

colonies may last for many years and grow very large, but their size is quite often restricted by swarming. Every now and then groups of females and workers leave the parental nest and go off to start a new one.

Wasps with a "peck order"

It is generally accepted that the multi-queen colonies of the poly-biine wasps are more primitive than the single-queen colonies that we find among the wasps of temperate regions. It also seems likely that the single-queen society has evolved from the multi-queen society as an adaption to the cooler and more seasonal climate of the temperate regions. The paper wasps of the genus *Polistes* show some interesting habits which support this idea. They are in many ways intermediate between the polybiine wasps and the familiar genus *Vespula*—the common wasps, or yellow-jackets, of Eurasia and North America.

Paper wasps make rather small nests consisting of a single layer of cells and hanging from trees or any other convenient support. In the tropical regions the colonies go on from year to year and they often contain several queens. One of the queens dominates the others, however, and there is a definite rank order among them, with the subordinate queens giving food to those above them in the hierarchy. The dominant queen does most of the egg-laying and may even destroy the eggs laid by the others. These tropical colonies often send out swarms consisting of one or more queens and a number of workers.

Polistes gallicus is a very common species in the southern half of Europe. The colonies are only annual affairs and, as with the genus *Vespula*, only the mated queens

Above: A complete Vespula nest.

Left: A paper wasp.

survive the winter. New nests are started in the spring and in the more southerly regions the nests are nearly always started by several queens working together. L. Pardi, an Italian zoologist, describes how one of these queens then asserts her dominance by attacking the others and emitting a loud buzzing noise. The subordinate or auxiliary queens do not do much egg-laying and generally disappear when the first new wasps begin to emerge in the colony. In the more northerly parts of the insect's range the queens never combine to form colonies: each one works alone just like the *Vespula* queen.

The first wasps to emerge in the new colony are all workers and they are considerably smaller than the queens. They help to rear further broods throughout the summer, and it is noticeable that successive broods produce larger and larger workers until, in late summer, there is no obvious difference between the workers and the queens. Male wasps then start to appear, followed by new queens. These pair up and the colony then disintegrates, leaving only the mated queens to survive

A mason wasp (Odynerus) entering its nest through the typical mud tube entrance.

the winter and begin the cycle again in the spring.

Yellow-jackets

Social life reaches its highest development in the wasps of the genus *Vespula*, commonly known as yellow-jackets in the United States. These are the yellow and black insects that make a nuisance of themselves in our houses during the late summer. Together with the closely related hornet, a larger and browner insect, they are found throughout the temperate regions of the northern hemisphere. Several species have been introduced into Australia and New Zealand, and a few species live in the tropical regions.

The tropical species form perennial colonies and periodically give off swarms of queens and workers, but the majority of the species form only annual colonies, founded in the spring by a single queen. On emerging from hibernation the queen begins to look for a suitable nesting site. Some species suspend their nests from trees and bushes, some build theirs inside hollow trees, and some nest under the ground.

The paper is first used to make a little stalk hanging from a branch or from the roof of the nesting cavity. The queen then makes a little cell at the bottom of the stalk and lays her first egg in it. Other cells are added around the first one until there is a little disk of perhaps thirty cells, all opening downwards and looking like a small *Polistes* nest except that the queen has provided an "umbrella" of overlapping paper plates. Eggs are laid in the cells before they are finished because it is important for the queen to get her family going as quickly as possible. The queen can look after only a certain number of grubs by herself and that is why she does not make very many cells. When her eggs hatch she begins to feed the larvae on chewed-up insects. The wasp grubs grow rapidly and soon turn into adult wasps. They are all workers and they set about enlarging the nest by making more layers of cells and covering them with more sheets of wasp paper. Those species nesting in the ground also have to dig out more soil to accommodate the growing nest.

When the first workers appear the queen stops work and devotes the rest of her life to laying eggs in the cells. She may lay more than 30,000 eggs during the year, although the colony never contains that many wasps at any one time because the individual workers live for only a few weeks and are continually being replaced by their younger sisters. The first workers are very small, but workers emerging later in the year are somewhat larger. This is probably the result of better feeding, since there are proportionately more workers to look after the later grubs. The workers are always smaller than the queen, however, and they differ in certain structural features as well. Although they can lay eggs, and many of them do later in the season, the presence of an active queen normally prevents them from doing so by inhibiting the development of their ovaries.

In the middle of the summer, when the colony has reached its peak population and the worker/larva ratio is at its highest, the wasps start to rear new queens instead of workers. Male wasps also make an appearance at about this time. Males come from unfertilized eggs, and although most of these eggs are produced by the queen, some of them undoubtedly come from the workers. The new queens pair up with the males and receive enough sperm to last them through the

Swarm queen cells of the honey bee (Apis mellifera), built on the edge of the comb.

following year. The new queens then seek out a suitable place for hibernation. The old workers, with no more brood to look after, have no need to search for insects any more and they embark on a prolonged feast of fruit and other sweet substances. It is at this time that they really come to our notice as they swarm around cake shops and kitchens, but we ought not to mind too much. During the previous few months these wasps and their elder sisters will have destroyed vast numbers of insect pests. The worker wasps, together with the old queen and any surviving males, die in the autumn when the frosts come, leaving only the new queens to continue the species in the following year.

Solitary and social bees

It is generally accepted that the bees evolved from some kind of solitary wasp about 100 million years ago, at the time when the flowering plants were evolving rapidly. Many bees are still remarkably wasp-like in appearance, but the essential difference between the two groups is that the bees feed their young on pollen and nectar instead of on the animal material used by the wasps. Most bees have evolved special structures for collecting pollen and nectar. Their body hairs, for example, are normally branched and well suited to trapping pollen. Hairs on certain regions, such as the legs, form "pollen baskets" in which the pollen is carried home. The bees' tongues are usually much more slender and longer than those of the wasps, allowing them to reach the nectar in the flowers.

One of the most primitive bees is a small black creature called *Prosopis*. There are actually several species and they are easily taken for small digger wasps. They are almost

hairless and they have evolved no special pollen-collecting apparatus. The pollen has to be swallowed together with the nectar. The tongue is still broad and wasp-like, so the insects can feed only at flowers, such as hogweed, where the nectar is freely exposed. *Prosopis* is a solitary bee and the female makes a small nest in the cavity of a reed stem. Each nest has perhaps half a dozen cells which the female stocks with a regurgitated mixture of pollen and nectar. An egg is laid in each cell and the cell is then closed. Having completed all her cells, the female has nothing more to do with them. She is probably dead before her eggs even hatch.

Bees of the genus *Halictus* exhibit a wide range of behaviour, from completely solitary to truly social. The best known of the truly social species is the European *H. malachurus*, studied by Fabre and many other entomologists. Mated females hibernate during the winter, often sharing a burrow with several others of their kind. In the spring the females—queens we can really call them—begin to fight and all but one of them leave the burrow to make their own homes. They usually start to dig quite near at hand and it is quite common to find a dozen nests in a square foot of ground. Occasionally two or more females nest so close to each other that they share a common entrance to their burrows. Each female digs down for three or four inches and she constructs about half a dozen little cells. As soon as each is complete she stocks it with food, lays an egg, and seals the cell. Such mass provisioning is unusual in an insect with social habits.

After a few weeks the first young bees emerge from the cells. They are smaller than their mother and they differ in their markings—so much so that they were originally thought to belong to another species. These new bees are the workers. They set about making more brood cells and they go out to collect nectar and pollen. The queen gives up her foraging trips and, when she is not laying eggs in the new cells, she acts as door-keeper. She does this by blocking the entrance with her head. Every now and then she will withdraw to allow one of the workers to go in or out, but she soon returns to her post. Towards the end of the summer each nest produces a few male bees and a few fully-developed females. These continue to live in the nest but they do not gather food while they are out and they contribute nothing towards the running of the colony. The old queen dies in the autumn and so do the workers and the males, leaving only the newly-mated queens to spend the winter in their burrows.

Bumble bees

The majority of bumble bees live in the temperate regions, both north and south, and their communities are only annual affairs. The mated female, or queen, hibernates underground and starts a new colony when she wakes in the spring.

Having selected her nest site, which may be under the ground or among grass tussocks, the queen makes a spherical chamber of the fine grass and other material she finds there, and then she goes out to collect pollen. She returns with the pollen bulging from the pollen baskets on her hind legs and deposits it on the floor of the nest, ramming it down firmly to make a bed. She is then ready to lay her eggs—in most species no more than about a dozen because she is unable to feed more than a small number of grubs. The eggs are laid on the bed of pollen. The queen's abdominal wax glands are active by now and she forms a wax cell around the eggs. In some species the wax cell is begun before the pollen is collected and the pollen is then deposited in it. Whatever happens, the eggs rest on pollen and are surrounded by walls of wax. The grubs hatching from the eggs

The bumble bee (Bombus) is a social insect but its colonies are not as large as those of the honey bee and they last for only one year.

begin to feed on the bed of pollen and the queen then provides more pollen and nectar as required. Most of this food is collected direct from the flowers, but the queen also has a store on which she can draw when the weather is bad—after laying her eggs she makes a wax pot and fills it with nectar.

When the first grubs have pupated the queen removes the wax coverings and uses the wax in making more egg cells on top of the cocoons. The first new bees emerge about a month after the eggs were laid—depending very much on the temperature and the species—and they very soon begin their duties of foraging for food and feeding the new larvae which are hatching at about this time. Once the first workers have appeared the queen gives up foraging, but she does not stop work altogether and she continues to help in the nest. In particular, the queen still retains the

The honey bee pupa lying in its cell.

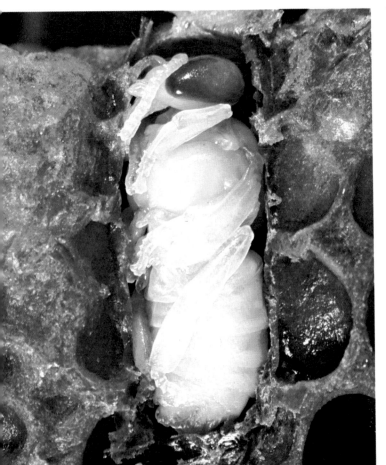

responsibility for making new egg cells. Each new batch is built on top of the previous batch of cocoons, and the number of new eggs laid each time depends upon the number of cocoons in the previous batch. The queen thus does not lay more eggs than can be managed by the workers. The old cocoons are often used for storing pollen and nectar.

Bumble bee colonies are never as large as those of the social wasps and they generally raise no more than a few hundred workers during the season. Species living in the colder regions—some of them reach far into the Arctic Circle—rear only a few dozen workers before the new males and females appear. This is obviously related to the short summer season at high latitudes.

New queens normally make their appearance when the colony has reached its maximum size. The old queen then begins to lay fewer eggs, and some of the eggs may well be eaten by the workers. The result is that the worker/larva ratio becomes large, with perhaps only one grub per worker. The larvae are very well fed under these conditions, and they develop into new queens. Male bees or drones generally appear a little before the new queens. They come from unfertilized eggs laid by the queen as a rule but occasionally by some of the workers. The males and the new queens pair up and the old colonies gradually break down. No more workers are produced and by the end of the summer most of the bees are dead. Only the new queens survive.

Cuckoo bees

Bumble bees, like wasps, are affected by social parasites. Many species of bumble bee are put upon by bees of the genus *Psithyrus*. The latter look very much like their hosts, except that they have no workers. The female *Psithyrus*—we cannot really call her a queen— hibernates like the *Bombus* queen but she does not usually wake up quite so early. When she does wake up her host species has already started a nest and the first workers are emerging. *Psithyrus* enters the nest and, if she overcomes the initial attack by the workers, she begins to lay her eggs. She usually kills the rightful queen and relies on the *Bombus* workers to rear her brood. No further workers are produced, of course, and the colony dies out much earlier than it would normally have done. The male and female *Psithyrus* bees pair up and the mated females then seek hibernation quarters.

The honey bee

There are four species of honey bee in the genus *Apis*, but only two have been "domesticated". These are the eastern honey bee (*A. indica*) and the western honey bee (*A. mellifera*). The two species are very similar in appearance and behaviour, and they have probably both descended fairly recently from a common ancestor somewhere in southern Asia. The natural home of these bees is a hollow tree, and so it has been quite easy for beekeepers to get them to nest in wooden hives and other artificial cavities. Both species are now reared all over the world and both have many varieties.

Honey bees have taken the social habit considerably further than the bumble bees and their colonies are perennial—indicating their tropical origin. There is a marked difference between the queen and her workers. She is considerably larger than the workers but her tongue is short and she lacks the pollen baskets on her hind legs. She also lacks the wax-making glands and the brood-food glands which will be mentioned later. She is quite unable to look after herself, let alone start and look after a colony. The honey bee colony is therefore always founded by a queen plus a group of workers.

The honey bee nest consists of a number of wax combs suspended vertically in the nesting chamber. Each comb consists of thousands of six-sided cells, all made by the workers with wax from their abdominal wax glands. There are two sizes of cell—worker cells which are about four millimetres in diameter, and drone cells which are slightly larger. The drone cells tend to be more common near the edges of the combs. Both kinds of cell are used for storing food as well as for rearing young bees.

As we have seen, the queen plays no part in building the combs. Her job is to lay eggs and at certain times of the year she produces more than a thousand eggs every day. She lays

The queen honey bee surrounded by attentive workers. She spends all her time laying eggs, but she also produces "queen substance" which is licked from her body by the workers.

an egg in each empty cell she finds and she leaves the workers to look after it. The egg hatches in a few days and the young grub is fed with brood-food, a protein-rich secretion from the salivary glands of the young workers. After three days on this diet the grub is switched to a diet of pollen and honey. A week later it is fully grown and ready to pupate. The workers seal the cell with wax, and after another week the young bee emerges.

Nearly all the new bees are workers and they soon take up their duties. The first few days are spent cleaning out the brood cells in readiness for the queen to lay more eggs. At this time the young bees also "beg" for food from their older sisters and they receive a regurgitated mixture of pollen and nectar. After about three days the young bees begin to help themselves to honey and pollen stored in the comb. The pollen is rich in protein and it assists the development of the brood-food glands. These start to secrete brood-food on about the sixth day of adult life, and the bee turns to the job of feeding the young larvae. This lasts for perhaps four or five days, after which time the bee is able to produce wax and she starts work as a builder. After a few days on this job the bee normally becomes a warehouse worker. She takes pollen and nectar from the returning foragers and carries it to the cells for storage. Then, after a series of exploratory flights, the bee embarks on the final phase of her life—foraging for food. Much of the information on this division of labour came from the German naturalist Dr G. A. Rösch in the 1920s, and he actually drew up a table showing how long each bee spends on each task. Later work, however—notably that by the naturalist Dr M. Lindauer—has shown that the bees' behaviour is more flexible than Rösch believed. Although most of the workers do each job in turn, some workers may miss out some tasks, depending on the needs of the colony. If all the bees of one age group are removed from the colony, their jobs will be performed quite efficiently by bees of another age group.

The dancing bees

The foraging bees collect both pollen and nectar, although each journey may be concerned with just one or the other. Pollen is collected on the body hairs and periodically brushed into the pollen baskets on the hind legs. When these are full the bee returns to the hive, often with some difficulty because the pollen weighs her down. An average honey bee colony is believed to collect up to a hundred pounds of pollen during a season. This means something like four million journeys for the bees— no wonder we call them busy bees.

Worker honey bees share the food they gather. Here, the bee with her mandibles open is giving food to the other two.

Honey bee larvae are all fed royal jelly or brood-food during the first three days of life. Thereafter, only the larvae destined to be queens receive it, the others being fed on a diet of honey and pollen.

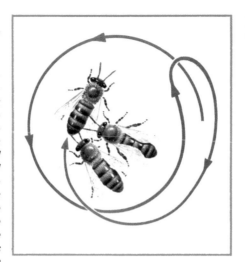

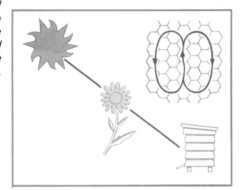

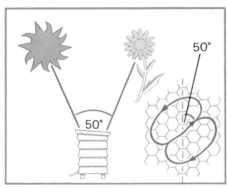

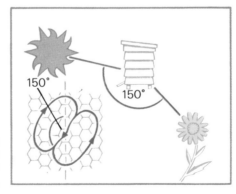

Much of the pollen brought back to the hive is eaten immediately, but the rest is stored in the cells. Nectar, which is basically a sugar solution, is swallowed by the bees in the field and regurgitated when they return to the hive. It is handed over to a younger bee which rolls the bead of nectar around in its mouth and evaporates much of the water from it. Chemical changes take place in the sugars and the nectar becomes converted into honey. Some of this is eaten right away, and the rest is stored in the cells.

The bees have evolved a wonderful way of telling each other about good sources of nectar and thus ensuring that they collect as much as possible for the hive. When a bee has found a good supply—a newly-opened patch of flowers, for example—it returns to the hive and "dances" on the comb. The Greeks knew about the dances more than two thousand years ago, but they had no idea why the bees danced. It was left to the brilliant German biologist Karl von Frisch to show that the dances inform the other bees of the exact whereabouts of the flowers. Von Frisch worked with bees for many years and he learned much about their ability to distinguish scents and colours, but even he was amazed by the results of his work on the bees' dances. Using glass-walled observation hives and saucers of syrup dotted about the surrounding countryside, he was able to show that the dances indicate both the direction and distance of the food from the hive.

Von Frisch described two dances which he called the *round dance* and the *waggle dance*. The round dance is performed by a bee which has found food within about fifteen yards of the hive. The bee returns to the comb and runs round in a circle, moving first this way and then

that. The other bees become very excited and they tag on behind the dancer, keeping in contact with their antennae. This chain of bees circles round a few times and then the leader breaks away and flies out of the hive. The other bees do not follow her immediately, but those which have been sufficiently stimulated soon fly out to search for the food. They remember the scent which the original forager brought back and they know that the food is not far away, so most of them soon find it.

When the food is more than about fifteen yards from the hive the bees perform a different dance. The circle opens out into two loops and, when the food is more than a hundred yards away, the dance consists of two complete semi-circles, rather like a figure-of-eight. This is the waggle dance, so called because the bee waggles her abdomen as she dances along the "straight run" in the centre. The waggle dance indicates both direction and distance of the food. When dancing on the comb, the bee uses the vertical to represent the direction of the sun, and the angle of the straight run in the centre of the dance then indicates the angle between the food and the sun. If the food lies 10 degrees to the right of the sun, the straight run will be 10 degrees to the right of the vertical; if the food is 70 degrees to the left of the sun the straight run will be 70 degrees to the left of the vertical; and if the bee's straight run is vertically downwards the food will be found in a direction completely opposite to that of the sun. The bees can measure the angles very accurately and von Frisch found that very few bees were more than 15 degrees out in the search.

The distance of the food is indicated by the speed of the dance— the nearer the food, the faster the dancer goes round. Von Frisch found that the bee made six straight runs in fifteen seconds when the food was five hundred metres away and only two runs in fifteen seconds when the food was five thousand metres away. It appears that the bees actually measure the distance by the amount of energy they expend getting to and from the food. Interesting results were obtained when the bees had to make a detour to avoid a high ridge. The hive was on one side of the ridge and the feeding place on the other, and von Frisch found that the bees' dances indicated the true direction of the food—the "bee-line" through the hill. The distance indicated by the dance, however, was the full distance around the hill—the distance the bees actually had to fly.

As well as their remarkable abilities to judge angles and distances, the bees have a wonderful sense of time. They are continually making allowances for the movement of the sun so that they can indicate the correct angles in their dances. Bees which are feeding in the evening due west of the hive—in the direction of the sun—make for the same place without any hesitation the next morning, although the sun is now in quite the reverse direction. Their time sense thus continues to work throughout the night.

The bees also use their "language" when they are searching for new nesting sites.

Queen substance

Wherever she goes in the hive, the queen bee is always surrounded by a retinue of workers who feed her and lick her body. While doing this they pick up the queen substance. This is a mixture of at least two substances which are secreted by glands on the queen's body. The workers gradually spread it through

New honey bee colonies are produced by swarming, with a queen and numerous workers going off to find a new home. The swarm hangs on a branch while scouts search for suitable quarters. Returning scouts perform a waggle dance on the surface of the swarm. Other bees then go to look at the sites and they eventually decide which one to adopt. The old colony continues to survive with a new queen as ruler. This new queen was already being reared by the workers before the old queen left with the swarm.

drone
or male

queen

worker

The three castes of the honey bee—note the bulging pollen baskets of the worker.

the summer. They do no work and they spend most of their time sitting on the comb or taking nectar from the cells. The workers throw them out of the hive in the autumn.

When the queen is getting old or unfit, her queen substance production slows down and the workers do not therefore obtain enough to keep them working normally. Within a few hours of the fall-off in queen substance the workers start to build queen cells. These are large thimble-shaped cells at the edge of the comb. The queen lays her eggs in these cells in the normal way, but the workers give them special treatment. Whereas the grubs in worker cells receive brood-food or royal jelly only during their first three days, the grubs in the queen cells get this food throughout their lives. It is this better diet which causes the grubs to grow into fully developed females or queens. The first new queen to emerge from her cell usually kills the other developing queens who would be her rivals, and then she goes off on her marriage flight. While she is away she usually mates with four or five drones and receives enough sperm to fertilize the 500,000 or more eggs she will lay in the next three or four years. She then returns to her hive and begins to lay, often side by side with her ailing mother—the only time when two queens are in one colony.

Colonies which have lost their queen—usually as a result of the beekeeper's experiments—immediately set about rearing an emergency queen. One or more worker cells in which there is an egg or a young grub are enlarged into queen cells and the queens are then reared in the normal way. This can be done only if there are grubs less than three days old. If there are no such grubs which can be brought up as queens the colony is doomed.

the whole population by their food-sharing activities and it is responsible for maintaining the smooth running of the colony. Queen substance seems to keep the workers happy and they become very agitated if the queen is removed. In particular, the queen substance prevents the development of the workers' ovaries and the formation of new queen cells. It also attracts drones when the queen is outside her hive.

The drone and the new queen

The drones arise from unfertilized eggs which the queen lays in the drone cells. The eggs and the resulting larvae are treated in exactly the same way as those in worker cells but, being unfertilized, they develop into males or drones. Each drone lives for about a month and there are probably several hundred of them in the hive at any one time during the summer, but they are most numerous towards the end of

A worker hive bee— a hive may contain as many as 80,000 bees, the vast majority being workers.

CHAPTER 7

THE ANTS

The ants, like the wasps and bees, belong to the large order of insects called the Hymenoptera. Their colonies, like those of the wasps and bees, consist mainly of sterile females or workers and they are "ruled" by one or more queens. There is, however, a much greater difference between the ant queen and her workers than we have seen between the queens and workers of the other social insects. The queen ant is usually a good deal larger than the workers and she normally has wings when she starts her adult life. Her eyes are usually much larger than those of the workers, many of whom are quite blind. The workers, which are all wingless, exhibit a wide variation in size even in one species, and many species possess an extra large worker caste which is known as a soldier. These have particularly large heads and jaws and their main job is to defend the colony. They use their large heads to block up holes in the wall until the smaller workers can repair them.

Male ants are found only at certain times of the year. They have wings, but they are generally somewhat smaller than the queens and they are very short-lived. In common with the males of the bees and wasps, they do nothing towards the upkeep of the colony.

Ant nests are far less elaborate buildings than those of the bees and wasps, although they may be much larger. Most of the ants nest under the ground, and the nest consists of a maze of tunnels and chambers. Some of the chambers are used as nurseries, others as store-houses, and some are even used as cemeteries. The soil excavated by the ants is often heaped up to form the familiar ant-hill, and the network of passageways then extends into the heap as well. The European wood ant (*Formica rufa*) builds great mounds of pine needles and small twigs, often ten feet across and a yard high. Unlike the pine needles on the surrounding forest floor, those of the ant-hill remain clean and untouched by fungi. Careful observation has revealed that the ants are forever working over the heap, bringing needles up to the surface and thus burying those formerly exposed to the air. Every twig and pine needle is moved in this way and it seems probable that, like bees, the ants produce anti-fungal material which prevents the nest materials from going mouldy. A number of ants make their nests from fragments of wood cemented together with saliva. The material thus produced is called carton, and it resembles a rather tough wasp paper. These carton nests are often constructed in hollow trees, and they sometimes hang from branches very much like wasp nests, but they contain nothing com-

Driver ants "march" in long columns and devour any animal material, dead or alive, which cannot get out of their way. These ants have no permanent homes.

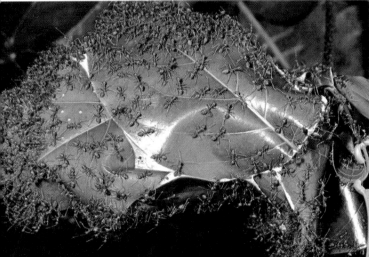

parable to the cells of the wasp nest.

Perhaps the strangest ant nests are those of the weaver ants, which belong to the genus *Oecophylla* and are found throughout the tropical regions of the Old World. These ants construct their nests in trees and shrubs by joining a number of leaves together with silk. The ants themselves cannot produce silk and it was some time before the source of their silk was discovered. The secret can be revealed by forcing open a nest—a somewhat painful business because the ants bite fiercely when disturbed—and watching the repairs which begin almost at once. A group of workers line up along one broken edge and reach out towards the other edge. If they can reach it with their jaws they pull it towards them and bring the two edges together. If the ants cannot reach the other edge a second batch of workers comes into operation and they all form chains in an attempt to bridge the gap. Each chain may need several ants, but eventually they manage to bring the two leaf edges together. More workers then appear, carrying larvae. Although the adult ants cannot make silk, the larvae can and they are employed as "shuttles" to weave a silken bond between the leaves. The workers dab the larvae down first on one side of the join and then on the other, attaching the sticky silk to both leaves. The silk soon hardens and holds the leaves firmly together.

Hunters and farmers

There are probably about 10,000 different species of ants and they are all social creatures—unlike the wasps and bees, of which only a small percentage lead social lives. We cannot, therefore, describe a sequence from solitary to social

species as we have done with the wasps and bees. We can, however, recognize several levels of ant civilization. The most primitive ants are hunters, living entirely by capturing other animals. Then there come the so-called food-gatherers which collect both plant and animal food. At the top of the scale we find the farmers which actually grow their own crops. William Wheeler, renowned for his studies of various social insects, has pointed out that this sequence is very similar to that found in human civilization.

The hunters and the farmers of the ant world live mainly in the tropical regions, leaving the temperate regions, especially those of the Northern Hemisphere, populated by the food-gathering ants. These are a rather mixed bunch, using a wide variety of materials for food. Most of them will eat animal food,

mainly in the form of other insects, but some of them rarely eat anything but seeds. The seed-eaters, commonly known as harvester ants, inhabit fairly dry regions and are very common in the Mediterranean area. Their nests are quite easy to find because the foragers go out in parties and they make distinct paths through the surrounding vegetation as they pass to and fro. They feed largely on grass seeds and they perform what seem to be very intelligent actions, although they are, of course, guided almost entirely by instinct. The seeds are stored in the chambers of the nest—food storage itself being an unusual habit for ants—and the ants take great care to maintain them in good condition. If the seeds get too damp the workers bring them out into the air and dry them. At least some of the species also bite out the "germ" of the seed

Far left: Weaver ants make their nests by "sewing" leaves together with silk produced by the grubs (top). The grubs are, in fact, used like shuttles and they pay out silk as they go.
The wood ant constructs a large conical nest which it covers with a thick layer of pine needles (bottom).

Below: Wood ants dragging a butterfly back to their nest.

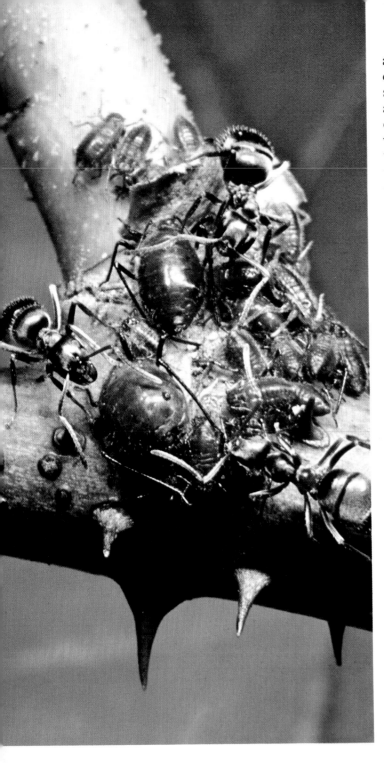

are extremely fond of the honey-dew excreted by aphids and other sap-sucking insects. If you look at an aphid-covered stem you will often see ants stroking the aphids with their antennae and nuzzling them with their jaws. This treatment causes the aphids to give up a drop of honey-dew, which is little more than concentrated plant sap with a high sugar content. So fond are the ants of this substance that many of them herd the aphids rather as we herd cattle, and Linnaeus actually referred to the aphids as "ant cows". The ants carry the aphids from plant to plant and even carry them into their nests during bad weather. Aphid eggs are often collected by ants during the autumn and stored in their nests before being put out on the plants again in the spring. Many ant species build protective tents for the aphids and stand guard over them, ready to bite or sting anything that tries to steal their charges.

Some ants are real pests. The carpenter ants of the Americas do almost as much damage to timber as the termites, while the leaf-cutting ants cause widespread destruction to plantations in South and Central America. The leaf-cutters belong to the tribe of ants called the Attini, and they represent the most advanced level of any civilization—the farming community. These ants feed entirely on fungi which they grow in large beds of chewed leaves. Long processions of workers can be seen wending their way back to the nest, each worker carrying a piece of leaf above her head. This habit has led to the common name of parasol ants. On arriving at the nest, the leaf fragments are chewed up and added to the underground mushroom fields. Vast numbers of leaves are used each day, hence the damage to plantations. A young queen

Many ants "milk" aphids for the sweet honey dew which they exude.

and so prevent it from germinating. The large-jawed soldiers find themselves with a new job among the harvesters: they use their powerful jaws to crack the seeds.

All the food-gathering ants like sweet substances, and many of them

leaving to begin a new colony has one very important piece of "luggage": she must always take a small piece of the fungus bed with her. Without it she will be unable to grow any food in her new home.

But ants are not entirely bad. They destroy an immense number of plant-eating insects, especially in the woodlands. Foresters like to have plenty of wood ants in their plantations and there is actually a trade in these creatures: nests are excavated and transported to new areas and even to other countries.

The ant's life story

Most people are familiar with the "plagues" of flying ants which occur during the summer. These are the "marriage flights" of the ants and it is convenient to take up the story at this point. The flying ants are all males and fully developed females and they pair up while in flight. Each pair soon falls to the ground and the insects separate. During this brief mating, however, the female receives enough sperm to last for the rest of her life—possibly ten years or more, during which time she will lay several million eggs. The male perishes very soon after mating, but the female's life is only just beginning. Her first task is to break off her wings by rubbing them against a stone or some other object, for her wings will be of no further use to her in the confines of the nest. The

Eggs, larvae and pupae are kept in different chambers of the ant nest, and the workers carry the young ants from place to place as they grow.

future development of this new queen depends largely on her species, for ants vary a good deal in the way in which they start new colonies. The following account deals specifically with a little black ant called *Lasius niger*. It is one of the commonest European species and very often nests under garden paths. Many similar species are found throughout the northern temperate regions.

When she has removed her wings the new queen may enter an established nest of her own species—possibly the nest in which she grew up. She is usually welcomed there because, unlike the majority of wasps and bees, the ants are quite happy to have more than one queen in the nest. Many species always have several queens in one nest, but *Lasius niger* is not one of these and the new queen usually starts her own colony. She finds a suitable hole and hides herself away for several months while her eggs mature inside her body. She does not feed at all during this period and she is sustained by the material derived from her degenerating flight muscles. Eventually she lays her first batch of eggs, but she still does not go out to collect food and when the eggs hatch she feeds the grubs with her own saliva. In a few weeks the grubs produce very small workers and these then set about the task of constructing the nest, feeding the near-starving queen, and bringing up the next batch of grubs. The queen does little work now, apart from laying eggs. She spends all her time in her "royal" chamber, with numerous workers in constant attendance. Her eggs are removed as soon as they are laid and they are put into special egg-chambers. Separate chambers are used for housing the grubs and the pupae.

Thousands of workers are reared and then, when the colony is mature, the reproductive castes are produced and another marriage flight takes place. The timing of the flight seems to depend on the weather and the workers will not allow the winged ants to emerge until conditions are just right. Then, within a very short space of time, all the nests in the neighbourhood release their flying ants. This synchronization means that there is a very good chance of a queen mating with a male from another colony and not with one of her own brothers. The chances of cross-breeding are often greatly enhanced by the fact that some colonies tend to produce mainly females and some tend to produce mainly males.

The ant colonies do not perish after sending out their mating swarms, for the queens are long-lived insects and their colonies go on from year to year. In theory those colonies which welcome new queens every now and then are immortal, although there is, of course, a fairly rapid turnover of

Only the fully developed male and female ants have wings. They appear only at certain times of the year, and the workers allow them to leave on their marriage flights only when the weather is right. We are then plagued with flying ants for a short time.

workers. Colonies living in the cooler regions hibernate in the depths of their nests during the winter.

Male ants are produced from unfertilized eggs, but we do not know how or under what conditions the queen lays these eggs. Nor do we know very much about the way in which the workers rear new queens, although a British biologist, Dr M. V. Brian, has made a detailed study of caste determination in *Myrmica rubra*. Whether an egg develops into a queen or a worker seems to depend on several factors, such as the size of egg, the age of the queen, the temperature, the food provided, and the amount of pheromone produced by the parent queen.

Ant armies

Hunting ants fall into two major groups—the ponerines and the dorylines. The ponerine ants, which live mainly in the Australasian region, form small colonies with no more than a few hundred individuals and they hunt singly. The dorylines, on the other hand, hunt in huge armies containing millions of individuals. These are the famous driver ants of Africa and the legionary ants of South and Central America. Both groups are sometimes called army ants.

The colonies contain workers of two sizes and soldiers which are about half an inch long. The soldiers have massive jaws and a tenacious bite, which African natives put to good use for "stitching" wounds. The edges of the wound are brought together and the soldier ants are made to bite into them. The ants are then beheaded, but their jaws remain firmly closed across the wound and they are left there until the wound has healed. The queen ant is a bloated sausage-like creature about

two inches long. She is nearly all abdomen and her head and thorax look very much like shrivelled pieces of skin at the front end. She never has any wings and, like the workers and soldiers, she is totally blind. Her mate is quite different: he is about the same size but has a much larger head and thorax, with large eyes and strong wings, and looks more like a wasp than an ant.

Army ants make no nests and have no permanent homes. They are generally active at night and spend the daytime hiding under fallen trees and so on, although many of the legionaries move about during the day. When they are on the move they form huge columns, hundreds of yards long and often many yards wide. The queen is half carried and half dragged along by the workers, while the soldiers march along on the outside of the column. The

Wood ants communicating by rubbing antennae.

occasional males tower above the column rather like stilt walkers tower above a circus procession.

Drivers and legionaries are entirely carnivorous and they eat any animal, dead or alive, which they find in their path. A column of driver ants has been known to kill and eat a whole caged leopard in one night, but large animals can normally get out of the way and the ants usually feed on smaller creatures. Nothing short of a wall of fire will divert the ants from their path and if they are making for a house the best thing to do is to get out and wait until they have passed through. Terrifying as it might be, such a visit is not altogether a bad thing: the ants are better than any insecticide or rat poison and, apart from leaving a few bones, they will remove all trace of pests from the house.

The queen has a definite egg-laying cycle and when she is about to lay a new batch of eggs the colony settles down in some place which will be home for the next three weeks or so. Raiding parties still go out each night but they return after each raid. The queen lays several thousand eggs while she is confined to base and when they hatch the colony becomes nomadic again, sleeping in a different place each day. The larvae are carried everywhere by the workers and fed by them until they are ready to pupate. By this time the queen is ready to lay more eggs and so the colony settles down again. The ants' nomadic wanderings are thus confined to the periods when they have to provide food for growing larvae.

The queens cannot go off on a marriage flight like other ant queens, and the army ants produce new colonies by dividing the parent colonies, rather like the swarming behaviour of the honey bee. The parent colony begins to split into two parts, one with the old queen and one with the larvae. One of the larvae is chosen for rearing into a new queen and the two parts of the colony become more and more distinct. Ants continue to pass from one part to the other for a time, but this traffic gradually dies out and the two groups become separate colonies. The new queen matures and mates with one of the many males produced at this time.

Parasites and slave-makers

Starting a new colony is a very hazardous business for a young queen. Large numbers die of starvation before they manage to raise their first worker daughters. It is not surprising, therefore, to find that several species have evolved ways of overcoming this problem. Some newly mated queens enter existing nests of their own kind and then, together with a group of workers, each queen establishes a satellite nest which often remains in contact with the parent nest. Some new queens depend on other species to help them and they become temporary social parasites. The American *Formica exsectoides*, for example, enters the nest of the closely related *Formica fusca* and begins to lay her eggs. These are tended by the host workers and, before long, the host queen is killed. No more host workers are produced and the original *fusca* colony is transformed into a pure *exsectoides* colony.

These temporary social parasites have their own workers, and once their colonies are established most of them are able to look after themselves. Some of them, however, employ "slaves" which they find by raiding the nests of other species. The raiders actually bring back pupae from the plundered nests, and the workers that result from these pupae set to work just as they would have done in their own nests.

Not all slave-makers start out as social parasites, however. *Formica sanguinea*, a red ant common in Europe and North America, is perfectly capable of starting its own colony and maintaining itself by its own efforts, and yet it very often engages in slave raids against closely related species. The amazon ants of the genus *Polyergus* are completely dependent upon slave labour for building their nests and feeding their young. The first small workers produced by the queen go off on a raid and no building is done until the stolen pupae produce workers.

Anergates atratulus, a rather rare European ant, has taken things even further. It is a complete parasite and, like the cuckoo wasps and bees, it has no workers. The mated queen enters the nest of the turf ant, *Tetramorium caespitum*, and lays her eggs there. Her presence somehow induces the *Tetramorium* workers to kill their own queen and they then devote themselves to rearing the young *Anergates* ants. The *Anergates* males are wingless, pupa-like creatures and they never leave the nest. They mate there and the females fly off to find other turf ant nests.

Invited and uninvited guests

Ant nests provide homes for many creatures besides the ants. The same is true of the nests of bees and wasps, but the ants' guests far outnumber those of the bees and wasps. It has been estimated that some 5,000 different kinds of animals have taken up residence in ant nests of one kind or another. Most of them are insects, especially beetles, but there are also numerous mites and small blind woodlice. Many of the guests are unwelcome because they steal the ants' food or attack the ant grubs, so they are attacked by the ants. The majority of the guests are harmless, however. They often do good by eating the rubbish that accumulates in the nest, so they are generally ignored by the ants. Then there are the welcome guests—the ones which the ants like to have around because they provide the sweet secretions so loved by most ants. Some are actually brought into the nest and cared for by the ants, who even go so far as to feed the guests with eggs and young ants. The caterpillar of the European large blue butterfly is one of the most famous of these invited guests. It spends the early part of its life feeding on wild thyme, but then it starts to produce sweet secretions and it attracts the attention of the ants. It is then carried into the nest and remains there until it changes into the adult butterfly.

Wood ants on the surface of their nest. There may be hundreds of thousands of ants in a single nest.

CHAPTER 8

THE TERMITES

All the social insects described so far belong to the Hymenoptera, one of the most advanced groups of insects. The termites, however, belong to the Isoptera—a much more primitive group which has strong links with the cockroaches. There are, nevertheless, some remarkable similarities between the termite societies and those of the ants. Termites are very often called white ants.

The termites are essentially tropical insects, although a few species do reach southern Europe and Canada. They are best known for their wood-eating habits, which result in enormous damage to buildings and to living trees as the insects chew their way through the timber. They will also destroy books and papers, and even plastics are not safe from them. But not all termites eat wood: the harvester termites feed mainly on grasses and they do much damage to crops. It has been estimated that termite damage to crops and buildings costs enormous amounts of money each year. Some of this loss is offset, perhaps, by the beneficial effects of some of the soil-living termites. Their tunnels help to drain and aerate the soil and thus promote good plant growth.

Compass termites

Termites make a wide variety of nests. Many of the more primitive species live in tunnels and chambers which they excavate in dead wood. They feed on the wood as well and they never need to leave their nests. Some termites excavate tunnels and chambers under the ground, but the most famous termite nests, or termitaria, are the huge mounds built by various members of the family Termitidae. They are composed partly of soil excavated from the subterranean tunnels and partly from material collected on the surface. The outer walls are extremely hard—sometimes as hard as concrete—but the bulk of the mound is composed of softer material in which the termites excavate numerous chambers. This inner material sets hard when it is mixed with water and it is sometimes used for making bricks. Nearly half a million bricks can be obtained from one large mound.

The shapes of the mounds vary from species to species, but one nest worthy of mention is that of the compass termite (*Omitermes meridionalis*), which lives in northern Australia. Each termitarium is rather like a wall, up to twelve feet high and perhaps twelve feet long but only two or three feet thick. The mounds are always built in a north to south direction and they can be very helpful to travellers. The value of this arrangement to the termites themselves is less obvious, but it probably exposes the nest walls to

Some of the more advanced termites construct mound nests far taller than a man. The nests are made of soil and the outer parts are cemented with saliva. They become as hard as concrete and are very difficult to remove. The inner regions are softer and they are traversed by numerous tunnels and chambers.

115

The opened nest of harvester termites, showing various chambers, the harvested grass and some nasute soldiers.

Each of the mound-building termites has its own design. These belong to the genus Cubitermes.

just the right amount of sunshine.

Apart from the various harvester species, which collect grass and other crops, the termites rarely come out into the open. They dig underground tunnels between their nests and their feeding grounds and they usually prefer to feed on wood which is partly buried. In this way they can enter the wood without coming above ground. If they do have to leave the soil to reach their food they often construct "footpaths" and completely roof them over with soil particles.

The royal couple

Whereas the ant colony is ruled by a queen and consists almost entirely of females, the termite colony is ruled jointly by a king and queen and the two sexes are represented more or less equally throughout the colony. The king and queen are the founders of the colony and they live for many years, often enclosed in a thick-walled "royal chamber". Unlike the queen ant, who mates only once in her life, the royal termites mate quite frequently throughout their lives. The queen's abdomen grows to an enormous size in some of the mound-building species and she can lay eggs at the rate of 1,500 an hour. The king and queen both possess wings early in their lives but, like the queen ant, they break their wings off when they begin to build the nest.

Juvenile workers

Ant workers are all adult females, but the termite workers are of both sexes and they are always young or juvenile insects. Young ants have no legs and they are quite helpless creatures. Young termites, on the other hand, have legs and antennae and they are quite able to do some of the work. Among the more primitive termite species the work is carried out by youngsters of various ages, but work is carried out only by the older nymphs in the more advanced species. The majority of the insects in the colony never progress beyond the worker stage, but they can become soldiers or reproductive forms if necessary.

Few termite species have been studied in any detail, but we have a good deal of information on the development of the castes of *Kalotermes flavicollis*, a rather primitive species found in southern Europe. The insects pass through four or five larval stages, during which they have no sign of wing buds. The amount of work done increases at each stage. Fourth and fifth stage larvae can, if necessary, moult and become soldiers. Fifth stage larvae can also moult and become replace-ment reproductives if the king or queen dies, but most of the fifth stage larvae go on to become pseudergates, or "false workers". These are the creatures that do most of the work. They still have no wing buds and the majority of them progress no further, but three other possibilities are open to them. They can become soldiers or replacement reproductives, or they can become nymphs. The latter have small wing buds, and under certain conditions they can develop into fully-winged adult termites.

The soldiers

The soldier caste is found in nearly all termite species and it generally accounts for about five per cent of the population. The main job of the soldiers is to defend the nest and its inhabitants. They appear very rapidly at the surface when a nest is disturbed and they attack the invaders with their powerful jaws. Some species have sharp edges to their jaws and they inflict deep wounds on their enemies, while other termite species possess "spring-loaded" jaws which they use to catapult away the invaders. A number of termites have gone in for chemical warfare: instead of having powerful jaws, the soldiers squirt a sticky fluid from a nozzle on the head. The

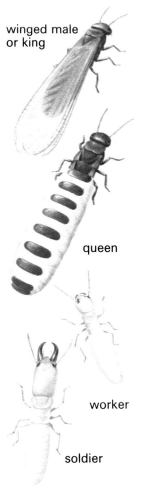

winged male or king

queen

worker

soldier

The four castes of Reticulitermes lucifugus. Note the remains of the wings on the queen and the swelling of her body caused by the eggs she is carrying.

Harvester termites carrying a grass stem back to their burrow.

Worker and soldier termites attending the queen whose swollen abdomen dwarfs them all.

A soldier termite defending a breach in the nest with its large jaws while workers repair the damage.

fluid traps any invading insects and it also gives off an alarm scent which warns the rest of the community. Soldiers of this type are called nasutes.

Termite soldiers are normally of both sexes, although one sex or the other may predominate in a given species. The soldiers, like the workers, are often of two or more distinct sizes. They also resemble the workers in being juvenile insects, but once they have become soldiers they cannot develop any further.

The importance of pheromones

We are only just beginning to understand what goes on inside a termite colony, but we can be quite sure that the fate of the young insects is controlled very largely by a complex system of pheromones acting in much the same way as the queen substance of the honey bee. In theory, young termites can all develop into reproductive forms, but the majority are prevented from doing so by the presence of the king and queen. It has been shown in a few species that the king produces a pheromone which acts upon glands in the young males and prevents them from becoming mature. The queen produces a pheromone which has the same effect on the young females. Removal of the royal couple results in the appearance of replacement or supplementary reproductives of both sexes. These develop from the workers and become sexually mature, although they have no proper wings. They take over the role of the king and queen. Numerous supplementary reproductives are produced, but they fight and, at least in *Kalotermes*, only one pair survives.

The soldier population is maintained at about the five per cent level by a similar system of pheromones.

When there are plenty of soldiers the pheromones they emit are sufficient to prevent the workers from developing into soldiers. But if the number of soldiers falls, or the number of workers rises, the concentration of the hormone falls in the colony and some of the workers begin to develop into soldiers. These pheromones are probably all distributed by way of food-sharing, or trophallaxis, which is very highly developed among the termites.

Pheromones are also of great importance to the termites when they are foraging. They lay down scent trails, usually with their abdomens, and other members of the colony can follow these trails to and from the nest. The members of each colony have a distinctive odour which enables them to recognize each other. They are aggressive towards members of other colonies of the same species, and each colony normally has its own territory. Members of other termite species are usually ignored.

The marriage flight

We do not know what causes the young termites to develop into fully-winged males and females, but at certain times of the year they do so and the winged termites, which are the new kings and queens, fly off in search of mates. The workers often build special passages through which the winged termites can reach the outside, but the flight does not take place until conditions are just right. In many species it occurs soon after the rainy season begins, and flights can be produced by watering a nest artificially. Pairing does not usually take place until the insects have finished their flights and they have got rid of their wings. They do this by rubbing them against a stone or by merely "shrugging their shoulders" and snapping the wings along a line of weakness. The females then raise their abdomens and emit a scent which attracts the males. A male grasps the female's abdomen and the pair move off in tandem for a while before digging a small hole in which they mate for the first time and rear their first brood of workers.

Some of the termites that live in wood and in underground galleries establish new colonies without sending out winged individuals. The outer galleries of these nests sometimes lose contact with the central regions and hormones from the royal pair do not get through. Supplementary reproductives are then produced in these outer galleries and they form distinct colonies.

The aftermath of the wedding flight—very few of the winged termites ever survive to start a new colony.

CHAPTER 9

THE COLONIAL ANIMALS

Careful examination of pond weed in a dish of water often reveals a small brown or green animal attached to the weed at one end and waving a bunch of slender tentacles at the other. This is *Hydra*, a little animal known to generations of biology students but rarely seen by anyone else. It is a relative, albeit a distant one, of the jellyfishes and the sea anemones and it uses its tentacles to catch the water fleas and other small creatures on which it feeds. The tentacles are armed with stinging cells, like miniature harpoons, which are discharged when an animal bumps into them.

When the water is warm enough *Hydra* reproduces itself by a process known as budding. One or more branches appear on the sides of the body and each develops a mouth and a ring of tentacles at its free end. The branches eventually break away and lead independent lives, but for a short while the parent and its offspring lead a colonial life, with two or more individuals joined together and sharing their food.

Budding is a very common method of reproduction among the coelenterates—the large group of animals to which *Hydra* belongs—and many species form permanent colonies. Here, the branches do not become separated from the parent animal, and when the branches themselves produce side-shoots they form a complex colony. Each member of the colony is called a polyp and, like those of *Hydra*, each one is hollow. The cavity of each polyp is continuous with that of its parent, and thus with that of any other member of the colony. Any food taken in by one of the polyps is therefore spread throughout the colony. We have, of course, seen food-sharing among the social insects, but here there is an even greater degree of interdependence—the individuals are physically joined to each other.

In the simplest colonies, such as the temporary colonies of *Hydra*, the polyps are all alike and all carry out the same jobs. Many colonial species, however, exhibit a division of labour within their colonies and they possess two or more types of polyp. *Obelia* is a very common colonial species which can usually be found growing on the larger seaweeds. The basal part of the colony consists of fine branching tubes cemented to the seaweed, and it is from these tubes that the polyp-bearing branches arise. The polyp-bearing branches stand up in the water like miniature trees and the animals are commonly known as sea firs—a name coined by the eighteenth-century naturalist John Ellis, who was the first man to study these little creatures. The polyps, each of which is seated in a horny cup, are of two types: feeding polyps,

Dendrophyllia, an attractive coral which grows on the Great Barrier Reef of Australia.

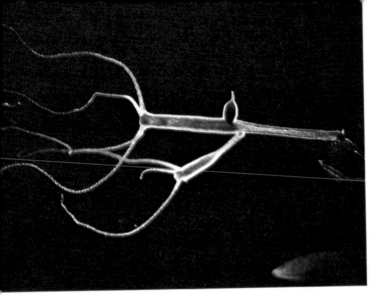

which resemble a rather squat *Hydra*, and reproductive polyps. The latter are pear-shaped and they have neither mouth nor tentacles.

Another well-known colonial animal is the Portuguese man-of-war. It is often called a jellyfish, but it is more closely related to *Hydra* and *Obelia* than to the true jellyfishes. The animal consists of a gas-filled float from which there hang several different kinds of polyp. Some of the polyps possess very long tentacles which hang down in the water rather like the drift nets used to catch herrings and other shoaling fishes. And the man-of-war's tentacles are just as efficient: when a fish bumps into them the stinging cells are discharged and the fish is trapped. The tentacles then contract and draw the fish up to the feeding polyps, which spread their mouths over it and begin to digest it. The Portuguese man-of-war is one of the few coelenterates whose stinging cells are dangerous to man.

The coral reef

The most famous of the colonial animals are undoubtedly the corals. We use this name for several groups of coelenterate animals which secrete limestone around themselves, but the various groups are not closely

related. The true corals, often called stony corals, are very closely akin to the sea anemones and each coral polyp looks very much like a small sea anemone. Many of the true corals are solitary creatures, each living in its own little cup, but the majority are colonial and in some parts of the ocean their colonies are so densely packed that they form huge banks or reefs.

The coral starts life as a small polyp, not unlike *Hydra* in appearance, although it has more tentacles and a more complex internal anatomy. Before long the animal secretes a little limestone platform under itself and then it begins to build up the sides of its cup, or theca, which will eventually surround it. The theca is more or less circular to start with but there are numerous radial walls growing inwards from the edge. These radial walls are responsible for the intricate patterns of many coral skeletons. Although the limestone is secreted on the outside of the animal, the coral always overflows its cup to some extent, so that it appears that the skeleton is inside the animal. The coral polyp branches in many different ways and this results in many different types of colony. Some are delicate branched structures in which the polyps are fairly well separated, while others are massive boulder-like structures with the polyps tightly packed together. The living corals bear little resemblance to the white skeletons that we see in museums because, when they are alive, the polyps form a continuous living film all over the skeleton and exhibit a wonderful variety of colours.

Reef-building corals grow only in the clear shallow waters of the tropical seas. Thousands of coral colonies of many different kinds grow up close together and, as polyp succeeds polyp, they build up layer upon layer of limestone. Fragments broken off by the waves or by browsing animals accumulate between the colonies and become cemented together by the action of various lime-secreting algae. A solid mass of limestone thus develops, covered by a living film. It is hard to realize that these reefs, some of them hundreds of miles long, have been formed by tiny animals rarely any larger than a pea. As well as the true corals, the reefs contain many other animals. Soft corals and other colonial coelenterates abound, and many fishes make their homes among the crevices. The coral reef is thus a very complex community, rivalling a tropical forest in the variety of living things making their homes there.

Moss animals

This rather contradictory name is applied to numerous aquatic animals, belonging to at least two different groups. They are also called bryozoans (the scientific way of saying "moss animals") or polyzoans. Most of them are colonial creatures and they feed by means of a number of small tentacles. The bodies of the moss animals, however, are much more complex than those of the corals. Some moss animals live in fresh water, but most of them are marine. Among the commonest are the sea mats, whose flat colonies cover rocks, shells and seaweeds. Other moss animals form delicate branching colonies, more like plants than animals. Most of the colonies consist of numerous horny or chalky "boxes", known as zooecia. These are rarely more than half a millimetre in diameter and each one normally houses an individual animal known as a zooid. Each zooid is distinct from its neighbours but there is sometimes a rather simple division of labour. Some zooids, for example, possess whip-like bristles and pincers which help to keep the whole colony free of dirt.

The sea mats — bryozoans whose colonies consist of horny little "boxes", each containing a tiny animal. The empty "boxes", shown here, are very commonly found attached to seaweed.

Staghorn (above) and brain coral (below), two rather different looking coral colonies produced as a result of different branching methods in the two types of animal.

The graptolites

As a final example of colonial creatures, we can look at the graptolites. More precisely, we can look at their fossils, for these little creatures became extinct some 350 million years ago. Graptolite fossils are very common in the fine-grained shale rocks that were laid down during the Ordovician Period, about 500 million years ago, and they usually appear like little white chalk marks. Looked at closely, they reveal a toothed structure, rather like tiny hacksaw blades, and it is believed that each "tooth" represents a cup in which there lived a tiny animal. If this is so, the animals were clearly colonial creatures, for there are usually a great many "teeth" on each stalk. Several species have branched stalks, some of them very much like tuning forks. It is thought that the graptolites floated freely in the sea because their fossils are widely scattered. Very little is known about the animals themselves, but they must have fed on small algae or other particles which they collected from the water.

Spreading the net

Colonial life has, as we have seen, evolved on several distinct occasions among the more lowly groups of animals. It obviously has some advantage over solitary life. The possibility of a division of labour is one advantage, although it has not been followed up by all of the colonial animals. The main advantage of colonial life, however, is in food gathering. All the colonial animals are sedentary or drifting creatures which rely on the currents to bring food, and it is obvious that a large branched colony has a much better chance of intercepting food than single animals have.

Conclusion

In this book we have looked at the levels of social life found among animals. Some of the advantages of group living have been pointed out, and we have seen how social life might have evolved from solitary habits by way of an extension of parental care and family life. Many mammalian societies have certainly evolved in this way, but true social life is less well developed among birds. We have therefore labelled most birds as sociable rather than social, indicating that they like to be together but do little in the way of helping each other. It is possible that the lack of social life among birds is associated with their inferior intelligence or learning ability when compared with mammals.

The latter part of the book deals mainly with the social insects— bees and wasps, ants and termites. These insects form huge societies with wonderful co-operation between the members. This does not mean that the animals are intelligent, but it shows their remarkable development of instinctive behaviour.

Finally, we have looked at the closest possible kind of co-operation, in which the individual animals are physically joined together. Such colonies can be formed only among the more lowly animals, but they came into existence in the same way as the social life of higher animals— through the increased and then permanent association of young animals with their parents.

"Tuning fork" graptolites (Didymograptus) fossilized on a piece of shale. These ancient colonial creatures died out more than 350 million years ago.

ACKNOWLEDGMENTS

Cover picture by Roger Tory Peterson, Bruce Coleman Ltd. Endpaper pictures by R. K. Murton and Norman Myers, Bruce Coleman Ltd. Helmut Albrecht, Bruce Coleman Ltd: p. 62. Anthony Bannister, Natural History Photographic Agency: pp. 12 (top), 20 (bottom), 78 (top), 82, 84, 85 (bottom), 86, 90, 116 (top), 118 (bottom). Des Bartlett, Bruce Coleman Ltd: p. 29. Jen and Des Bartlett, Bruce Coleman Ltd: pp. 22, 23 (left). F. Bel and G. Vienne, Jacana: p. 52 (top). Mark N. Boulton, Bruce Coleman Ltd: pp. 80–1. Jane Burton, Bruce Coleman Ltd: pp. 35 (top), 48, 51, 107, 113, 117 (bottom), 123 (bottom). P. J. K. Burton, Natural Science Photos: p. 15. Arthur Butler, Natural History Photographic Agency: p. 68 (top). C. G. Butler, Bruce Coleman Ltd: p. 96. N. A. Callow, Natural History Photographic Agency: p. 16. Bob Campbell, Bruce Coleman Ltd: p. 118 (top). Michael Chinery: pp. 11 (top and bottom), 73 (left), 85 (top), 87 (left), 88 (bottom), 125. Colour Library International: pp. 8, 10, 50. Stephen Dalton, A.I.I.P., F.R.P.S., Natural History Photographic Agency: pp. 37, 89 (bottom), 92, 93, 95, 97, 99, 100, 103, 106 (bottom), 111. A. J. Deane, Bruce Coleman Ltd: p. 74 (top). Pierre Dupont, Jacana: p. 26. Ake Wallentin Engman, Bruce Coleman Ltd: pp. 72, 75. M. Fievet, Jacana: p. 119. J. A. Grant, Natural Science Photos: pp. 21 (top), 25 (top), 108, 110. F. Greenaway, Natural History Photographic Agency: p. 20 (upper). Grossa, Jacana: pp. 25 (bottom), 65. E. H. Herbert, Natural Science Photos: p. 69. David C. Houston,

Bruce Coleman Ltd: pp. 6, 78 (bottom), 79, 80. Peter Jackson, Bruce Coleman Ltd: pp. 23 (right), 66, 114. Ernest James, Natural History Photographic Agency: p. 68 (bottom). R. S. L. Jonklaas, Natural History Photographic Agency: p. 38. Russ Kinne, Bruce Coleman Ltd: p. 35 (bottom). Geoffrey Kinns, Natural Science Photos: pp. 36, 40, 47. Yves Lanceau, Jacana: pp. 18, 94. F. Massart, Jacana: p. 46 (top). R. K. Murton, Bruce Coleman Ltd: p. 24. Norman Myers, Bruce Coleman Ltd: pp. 56–7, 104. Claude Nardin, Jacana: p. 39. Ernest Neal, Bruce Coleman Ltd: p. 31. Oxford Scientific Films, Bruce Coleman Ltd: p. 122. A. G. McR. Pearce, Bruce Coleman Ltd: p. 109. Graham Pizzey, Natural History Photographic Agency: pp. 76–7, 124 (top). Ivan Polunin, Natural History Photographic Agency: pp. 12 (lower), 106 (top and middle). Allan Power, Bruce Coleman Ltd: pp. 32, 120, 124 (bottom). Jacques Robert, Jacana: p. 73 (right). Michael J. Roberts, Barnaby's Picture Library: p. 17. Guy Roots, Barnaby's Picture Library: p. 60. W. A. Sands, Natural Science Photos: p. 116 (bottom). A. Shears, Natural Science Photos: p. 89 (top). James Simon, Bruce Coleman Ltd: pp. 43, 58. "Souricate", Jacana: p. 45 (top and bottom). David Stephen: pp. 42, 71. John X. Sundance, Jacana: p. 54. A. J. Sutcliffe, Natural Science Photos: pp. 49, 59. Barrie Thomas, Bruce Coleman Ltd: p. 44. Tollu, Jacana: p. 46 (bottom). M. W. F. Tweedie, Natural History Photographic Agency: p. 21 (bottom). David Urry, Bruce Coleman Ltd: p. 74 (bottom). Frans J. Vahrmeyer, Barnaby's Picture Library: p. 53. Verzier, Jacana: p. 55. P. H. Ward, Natural Science Photos: pp. 13 (upper), 21 (middle), 87 (right). P. Wayne, Natural History Photographic Agency: p. 27. Joe van Wormer, Bruce Coleman Ltd: p. 52 (bottom).

Drawings by Malcolm McLellan: pp. 13, 123 (top). Denis Ovendon: pp. 88, 98, 102, 117. Michael Hopkin, with the permission of B.P.C.: p. 34.

FURTHER READING

Butler, C. G. *The World of the Honeybee* Collins, London 1962

Chauvin, R. *Animal Societies* Gollancz, London 1968; Hill & Wang, New York 1968

Crook, J. H. *Social Behaviour in Birds and Mammals* Academic Press, New York 1970

Darling, F. F. *A Herd of Red Deer* Oxford University Press 1937

Davis, D. E. *Integral Animal Behaviour* Macmillan, London 1966

Fabre, J. H. *Social Life in the Insect World* Gale, New York 1912

Free, F. B. and Butler, C. G. *Bumblebees* Collins, London 1959

Frisch, K. von *The Dancing Bees* Methuen, London 1954; Harcourt Brace Jovanovich, New York 1965

Froman, R. *The Great Reaching Out: How Living Beings Communicate* World Publishing Company, New York 1968

Gilbert, B. *How Animals Communicate* Pantheon Books, New York 1966; Angus and Robertson, London 1967

Howse, P. E. *Termites: A Study in Social Behaviour* Hutchinson, London 1970

Lawick-Goodall, H. and J. van *Innocent Killers* Collins, London 1970; Houghton Mifflin, Boston 1971

Lawick-Goodall, J. van *In the Shadow of Man* Collins, London 1970; Houghton Mifflin, Boston 1971

Lockley, R. M. *The Private Life of the Rabbit* Deutsch, London 1964

Lorenz, K. *King Solomon's Ring* Thomas Y. Crowell, New York 1962; Methuen, London

Mason, G. F. *Animal Sounds* William Morrow, New York 1948; Dent, London 1959

Ribbands, C. R. *The Behaviour and Social Life of Honeybees* Bee Research Association, London 1953; Dover Publications, New York 1953

Richards, O. W. *The Social Insects* MacDonald, London 1953; Peter Smith, Gloucester, Mass.

Schaller, G. B. *The Mountain Gorilla* Chicago University Press 1964; Penguin Books, London 1967

Selsam, M. E. *The Courtship of Animals* William Morrow, New York 1964; World's Work, Tadworth 1970

Simon, H. *Partners, Guests, and Parasites; Coexistence in Nature* The Viking Press, New York 1970

Southwick, C. H. (Ed.) *Primate Social Behaviour* Van Nostrand Reinhold, London and New York 1963

Tinbergen, N. *Animal Behavior* Time Inc., New York 1965

Tinbergen, N. *Social Behaviour in Animals* Methuen, London 1964; Barnes & Noble, New York

Wheeler, W. M. *The Social Insects* Kegan Paul, London 1928

Note: Most libraries have sizable collections of books on specific species referred to in the text, many of which contain further information on families and communication of that species.

INDEX

Italic page numbers refer to illustrations